THE FUTURE STARTS HERE

THE FUTURE STARTS HERE

AN OPTIMISTIC GUIDE TO WHAT COMES NEXT

JOHN HIGGS

WEIDENFELD & NICOLSON

First published in Great Britain in 2019 by Weidenfeld & Nicolson
This paperback edition published in 2020 by Weidenfeld & Nicolson
an imprint of The Orion Publishing Group Ltd
Carmelite House, 50 Victoria Embankment
London EC4Y 0DZ

An Hachette UK Company

1 3 5 7 9 10 8 6 4 2

Text copyright © John Higgs 2019
Illustrations copyright © Melinda Gebbie 2019

A CIP catalogue record for this book is
available from the British Library.

ISBN (Mass Market Paperback) 978 1 4746 0940 1
ISBN (eBook) 978 1 4746 0941 8

Typeset at The Spartan Press Ltd,
Lymington, Hants

Printed and bound in Great Britain by Clays Ltd,
Elcograf S.p.A.

MIX
Paper from
responsible sources
FSC® C104740

www.orionbooks.co.uk
www.weidenfeldandnicolson.co.uk

CONTENTS

'Where there is no vision, the people perish.'
Proverbs 29:18

'We're all in the gutter, but some of us are looking at the kerb.'
Joss Cope

INTRODUCTION:
THE COMEDY OF ERAS

1.

At some point in the 1980s we gave up on the future. Before then, we imagined wonderful days to come, free from disease, work and want. Thanks to television series like *Star Trek* or events like the 1939 Futurama World Fair, this shining vision became so ingrained in our culture that it could be spoofed by children's cartoons such as *The Jetsons* or *Futurama*.

At the Futurama World Fair, a time capsule was buried 50 feet under New York's Flushing Meadows. It contained items such as seeds, a pack of cigarettes, a department store catalogue on microfilm and a message from Albert Einstein. This capsule was designed to be opened in the year 6939. When we thought about the future in the 1930s, we assumed that our civilisation would be alive and thriving in 5,000 years' time.

Since then, our vision of the future has changed. When we look ahead now, we tell dystopian stories of environmental collapse, zombie plagues and the end of civilisation. Hollywood, TV and Young Adult novels all sing the same tune: it is downhill from here. If it is true that we must imagine the future before we can build it, then this is deeply worrying.

It is sobering to think back over all the films you've seen in the past few decades, and try to find a vision of the future that you'd want to live in. As far as I can tell, the last attempt at a utopian future in a mainstream Hollywood film was the 1989 comedy *Bill and Ted's Excellent Adventure*. This was a future described as being fairly similar to now, but with better waterslides. By 1989, that was the best future our imagination could offer.

When the post-apocalyptic Australian film *Mad Max 2* was produced in 1981, it was necessary to explain to the audience why civilisation had collapsed. Text at the start of the film explained how, in the near future, the world would run out of oil and this would lead to the dystopia depicted in the story that followed. Now, it is no longer necessary for writers and filmmakers to explain the post-collapse worlds they explore. There is no explanation about what went wrong in Cormac McCarthy's hope-killing 2006 novel *The Road*, nor is there any explanation for the existence of zombies in the long-running comic book series *The Walking Dead*. Audiences are primed to accept this narrative shorthand because dystopia, we understand, is what our future is going to be. P. D. James's 1992 novel *The Children of Men* describes a disintegrating British society in the year 2021 which is devoid of hope and purpose and increasingly bleak and violent. The reason given for this is unexplained mass sterility which caused children to stop being born in the mid-1990s. What's disturbing about the book now is that it feels like we are heading towards a similar society, and we don't need the plot device of mass sterility to get us there.

Even dystopias that were intended as jokes have come to seem horribly prophetic. In the same year as *Bill and Ted*, the film *Back to the Future II* took us forward 30 years and showed us a broken-down future which was violent, divided and ruled over by a moronic billionaire inspired by Donald Trump. As the director Robert Zemeckis said, 'rather than trying to make a scientifically sound prediction that we were probably going to get wrong anyway, we figured, let's just make it funny'. Thirty years later, the joke has become a little hollow. As the American writer Adam Sternbergh has noted, 'the biggest problem with imagining dystopia seems to be coming up with some future world that's worse than what's happening right now'.

If we judge by the stories we're fed by film and television, then our current civilisation can feel like a crime novel with the last page ripped out. We don't know exactly the identity of the murderer, but we do know that the story is about to come to an end. Perhaps a new antibiotic-resistant disease will erupt into a global pandemic and wipe us from the face of the Earth. Maybe a massive solar flare will destroy our electronic equipment and send us back to the Stone Age. There could be a killer robot uprising just around the corner, or perhaps our old favourite, nuclear war, will make a comeback. Even if we avoid all these possibilities, economic collapse caused by environmental devastation, climate change and the collapse of biodiversity is still getting closer, no matter how hard we try not to think about it.

In our current functioning society, an ATM will give me bank-notes that I can then use in the local supermarket, which has shelves piled high with whatever I need. As long as this situation continues, I know that civilisation is still working. We had a narrow escape during the financial crisis of 2008, when Britain was only two hours away from the ATMs being turned off. The switching-off of ATMs is, I suspect, a more likely scenario for the end than alien spaceships filling the sky or mushroom clouds over London. Visa card chip-and-pin payments failed across Europe for a few hours in June 2018, and over 5 million attempts to pay for everyday necessities such as petrol and groceries failed. Although this outage only affected Visa, it still demonstrated how reliant we have become on an electronic financial system, and how much we have moved away from physical money.

Perhaps a level in a videogame is a more apt metaphor for our current world than the end of a book. When something comes to an end in a story, it does so in a way that is meaningful in terms of the larger narrative. In a videogame, you keep going for as long as you can, but eventually you make one slip and then it's all over. There's no purpose or logic to such an end. It can come at any time. It does not matter how long you've been playing, or how brilliantly you've overcome the many problems encountered in the past. Any point could be Game Over, and there are no continues or extra lives.

I am old enough to remember when we still had a future. When I was growing up in the 1970s, the idea that society was progressing, and that life would get better, was still a part of mainstream culture. Politically, the 1970s were a time of economic stagnation, strikes and upheaval, but we could still imagine that these were temporary frustrations and that an exciting future remained just over the horizon. America landed on the moon six times between 1969 and 1972, so a future in outer space appeared to have begun. Miraculous-sounding machines called computers had the potential to supercharge our technology. Glam rock and musicians such as David Bowie suggested our golden future could be a dazzling time of freedom and liberation. The idea of constant progress and improvement had been the dominant Western narrative for centuries, embedded in everything from the Age of Enlightenment to the American Dream. When the Sex Pistols emerged like harbingers of an unwelcome nightmare and announced that we had 'no future', that was really shocking.

2.

In 2015, Disney produced a big-budget family adventure film about our lost dream of the future. As the director, Brad Bird, explained, 'When [co-writer Damon Lindelof and I] were little, people had a very positive idea about the future, even though there were bad things going on in the world. Even the 1964 World's Fair happened during the Cold War. But there was a sense we could overcome them. And yet now we act like we're passengers on a bus with no say in where it's going, with no realization that we collectively write the future every day and can make it so much better than it otherwise would be.'

Bird's film, which starred George Clooney and Hugh Laurie, took its name from the future-themed land in Disney theme parks: *Tomorrowland*. It turned out that audiences were not interested enough in the future to find out where it had disappeared to. The film was a box-office flop. It made a loss of over $100m.

In its review, the *New York Times* said that 'It's important to note that *Tomorrowland* is not disappointing in the usual way. It's not another glib, phoned-in piece of franchise mediocrity but rather a work of evident passion and conviction. What it isn't is in any way convincing or enchanting.' Bird and Lindelof knew that the future was a spell that used to be able to inspire wide-eyed wonder in us all. They also knew that this spell had stopped working. But they were not powerful enough storytellers to kick-start it back into life. The *Philadelphia Inquirer* review is probably sadly accurate when it said that the film is 'Unlikely to be remembered in decades to come – or even in months to come, once the next teenage dystopian fantasy inserts itself into movie houses.'

Had an audience turned up at the cinema, they would have been told that the *Jetsons*-style Utopia promised in the 1950s was

5

not our future. This Utopia existed instead in an alternative dimension, where it attempted to inspire us to build a paradise on Earth. When this didn't work, it began to warn us that we were heading for disaster. Unfortunately, this made things worse, as Hugh Laurie explains in a scenery-chewing speech near the end of the film. 'How do you think people responded to the prospect of imminent doom? They gobbled it up, like a chocolate éclair,' he declared, proving beyond doubt that he will never be short of work as long as Hollywood requires British villains who can monologue with style. 'They didn't fear their demise, they repackaged it. It can be enjoyed as videogames, as TV shows, books, movies. The entire world wholeheartedly embraced the apocalypse and sprinted towards it with gleeful abandon! [. . .] In every moment, there is the possibility of a better future, but you people won't believe it. And because you won't believe it, you won't do what is necessary to make it a reality. So you dwell on this terrible future, and you resign yourselves to it. For one reason – because that future doesn't ask anything of you today. So, yes, we saw the iceberg, we warned the *Titanic*. But you all just steered for it anyway, full steam ahead. Why? Because you want to sink. You gave up.' Laurie's speech is remarkable because it claims that the real villains in our story are the wider movie-going public. The audience are presented with an awkward truth instead of the usual comforting Hollywood praise. This was perhaps a factor in the film's failure at the box office.

Eventually, the film concludes that the real problem is the idea of Tomorrowland itself, full of jetpacks and scientific wonder, unreachable for us in its alternative dimension but still reminding us of what we are not. By its absence, Tomorrowland leaves us seeing ourselves as a failure. At this point, the film produces the traditional blockbuster movie mega-bomb, and uses it to blow up Tomorrowland itself (or, at least, its central tower). This, I think, is quite bold and insightful storytelling from Bird and Lindelof.

When you are trapped in a dark shadow, it is necessary to destroy that which casts the shadow. But this still leaves the problem of what to replace it with. This is where most reviewers thought that the film failed. At the end of *Tomorrowland*, Bird and Lindelof try to replace the previous, *Jetsons*-style future vision with nothing but vaguely defined optimism.

Throughout the film, characters reference a myth about two wolves. One wolf is darkness and despair, the other is light and hope, and they are always fighting. The question is, which will win? The answer, the film tells us, is whichever one you feed. The problem with this is that the light and hope that the film alludes to is vague and ill-defined, while the darkness and despair are very real. When the film depicts the future we're heading towards, it shows us floods, fires, melting glaciers, riots and the collapse of civilisation as we know it. These visions feel horribly plausible. They are supported by peer-reviewed scientific research. In contrast, the generalised faith in human creativity that the film ends on feels insubstantial and unconvincing. Choosing which wolf to feed will not select the winner when one wolf has been fed a hearty diet for decades and the other is a half-starved bag of bones.

'Perhaps *Tomorrowland* should not be blamed for succumbing to the poverty of vision that it works so hard to attack,' the *New York Times* review argued. 'Maybe the forces of negativity are just too strong. But it's also possible that the movie is confused about how to imagine and oppose those forces. False cheer can be just as insidious as easy despair. And the world hardly suffers from a shortage of empty encouragement, of sponsored inducements to emulate various dreamers and disrupters, of bland universal appeals to the power of individuality. *Tomorrowland* works entirely at that level, which is to say in the vocabulary of advertisement. Its idea of the future is abstract, theoretical and empty, and it can only fill in the blank space with exhortations to belief and to hope.

But belief without content, without a critical picture of the world as it is, is really just propaganda.'

The problem that defeated Bird and Lindelof has been weighing on my mind also. Is the reason why we fail to imagine a future simply that we genuinely don't have one, as climate science and rampant inequality suggest? Is a search for an alternative perspective doomed from the start? Is our oncoming end as certain as we are told?

3.

The future hasn't happened yet. The idea that our civilisation is doomed is not established fact. It is a story we tell ourselves. It is the latest in a very long line of stories.

Between 100,000 and 50,000 years ago, something about our ancestors changed. Anthropologists call this change 'behavioural modernity', and the Israeli historian Yuval Noah Harari calls it 'the cognitive revolution'. Either the arrival of language or a mental change that allowed the development of language took us out of the natural, animal world and made humans different from the rest of the fauna.

The earliest physical evidence we have of non-abstract human imagination was found in a cave near Stadel, Germany. It is a mammoth tusk which, 40,000 years ago, was carved into the figure of a human male with a lion's head. It was frequently handled, and the recipient of a great deal of effort and attention. It is estimated that, without blades or metal tools, it would have taken 400 hours to carve. The Stadel Lion Man is the first unreal, imagined thing to manifest in the archaeological record.

In the centuries that followed this original fantasy, things that don't physically exist exerted increasing influence over our lives.

Those immaterial things include spirits, gods, boundaries, tribes, laws, myths, flags, money and corporations. If these things have a physical element it is little more than a piece of cloth or paper, or a small chunk of earth. It is in their immaterial aspect, which we dream up and project onto the physical world, where their power and importance lie.

Modern people are defined by our fictions. When we go to work, or exchange wealth for goods, or enlist in wars against people we have never met, we do so because a web of shared stories gives those actions context. They allow us to synchronise our actions with millions of others. No other species can create the immaterial fictions needed to allow such collaboration, which is why no other species is able to co-operate on the scale that humans can. These narratives, which include the rules of our languages, the laws we obey and the entertainment we enjoy, structure the world in which we live. In the millennia since the carving of the Stadel Lion Man, we have moved out of the material world and into an immaterial one.

An overarching story which shapes a society emerges from the accumulation of countless smaller fictions. It was called the 'living mythology' by the American mythologist Joseph Campbell. The French philosopher Jean-François Lyotard called it our *grand narratif*. My preferred name for it is the 'circumambient mythos'. The word 'circumambient' suggests something that entirely surrounds us, but also permeates everything so completely that we almost don't notice its presence. The importance of the circumambient mythos is why I was talking about mainstream Hollywood movies instead of the more optimistic futures that can occasionally be found in sci-fi literature, such as the work of Iain M. Banks or in the Solarpunk or Afrofuturist movements. It is in the bigger picture that the wider concerns of our culture reveal themselves.

One reason for our obsession with dystopian futures is, I

suspect, the fact that the plot holes within our current circum-ambient mythos are increasingly difficult to ignore. An obvious example is the story of economic growth. Our economy is dependent on the economy continuing to grow and the future being wealthier than the past. Lending money, and expecting to receive a larger sum in return thanks to interest, can only occur when it is believed the economy of the future will be larger than the economy of the present. A glitch in this story was responsible for the credit crunch of 2008, when markets had something of a dark night of the soul and found themselves doubting whether that story was actually true. National governments printed an incredible amount of money to ease the markets from their doubts, but this is not something that we can afford to repeat.

It's not hard to see the flaws in stories like these. If the economy is to keep growing, we will have to consume more and more of the Earth's resources. However, we are already using more than our ecosystems can replace. Earth Overshoot Day is marked on the date when our resource use has exceeded what the Earth can produce in a year and, in 2018, this was 1 August. According to the UN, if current population and consumption trends continue, then by 2030 humanity will need the equivalent of two Earths to support ourselves. We don't have two Earths. As Stein's law tells us, 'If something cannot go on forever, it will stop.' It is worth remembering that Stein's law is often amended by Davies's corollary. This states that 'Things that can't go on forever, go on much longer than you think.'

According to the journalist Christopher Booker, the plots of all stories fall into seven basic categories. The names he gives to these seven plots are Overcoming the Monster, Rags to Riches, The Quest, Voyage and Return, Comedy, Tragedy, and Rebirth. The circumambient mythos, being a story itself, must always fall into one of these categories. In the medieval Christian West, for example, the overriding plot structure of our culture was Voyage and Return. We had come from God, and it was back to God we would return. This narrative explained a relatively stable society with little in the way of technological change or economic expansion, and it remained the plot of our culture for many centuries. If there was a better world somewhere, then this was understood to be the Classical civilisations of the past. The idea that a better earthly world resided in the future was not part of the story.

Around 500 years ago we slowly started to move to a different plot. As the insights of the Renaissance led to the Enlightenment, our focus moved from waiting for the afterlife to actively working to materially improve our time on Earth. Our story was now one of progress. This circumambient mythos is the reason why we automatically speak of 'technological advancement' instead of the more accurate 'technological change', even when we know that new technology does not necessarily make our society better. The name of this new plot, according to Booker's list, was The Quest. We were on a journey to a better place, provided we could overcome all the obstacles that the journey tests us with.

If Hollywood is to be believed, we've now stopped using the plot of The Quest, and the idea of progress, to understand ourselves. As the sociologist Robert Nisbet has written, 'The skepticism regarding Western progress that was once confined to a very small number of intellectuals in the nineteenth century has grown and spread to not merely the large majority of intellectuals in this final quarter of the [twentieth] century, but to many millions of other people in the West.'

The story structure which Western culture adopted to replace The Quest is Tragedy. Tragedy, Booker tells us, is the story form that always ends in defeat. According to Aristotle, the downfall of a character in a tragedy is not caused by outside forces, such as the gods or fate. Nor is it the result of vice or moral deficiency. Instead, there is a central character flaw in the heart of the hero which cannot be resolved. Aristotle used the word *hamartia* to describe this flaw, which translates as to miss the mark or to err. To possess *hamartia* is not to be a bad person, for there is no moral judgement involved. But it compels you to act in a way that causes events to evade your control, and these actions inevitably result in destruction.

Booker, in the spirit of literary theorists since Aristotle, defines Tragedy in a particular way. Tragedy 'shows a hero being tempted or impelled into a course of action which is in some way dark or forbidden', he wrote. 'For a time, as the hero embarks on a course, he enjoys almost unbelievable, dreamlike success,' he continued. 'But somehow it is in the nature of the course he is pursuing that he cannot achieve satisfaction. His mood is increasingly chequered by a sense of frustration. As he still pursues his dream, vainly trying to make his position secure, he begins to feel more and more threatened – things have got out of control. The original dream has soured into a nightmare and everything is going more and more wrong. This eventually culminates in the hero's violent destruction.' This is not, I think, a million miles away from how we see ourselves today.

But there is also a narrative plot which, for the characters living it, appears to be identical to tragedy. That plot is comedy. Nowadays, we think of comedy as something funny which has jokes in it. For this reason, schoolchildren often complain that the Greek or Elizabethan 'comedy' they study is not funny. But, technically, comedy is not defined by laughs. A comedy is a story that uses a plot structure similar to tragedy, except that the character flaw or *hamartia* at the heart of the story is not fatal. It can be resolved. In doing so this leads to a happy ending or a loving union.

Traditional comedies are about people who don't see themselves as who they truly are. They tell of peasants who have no idea that they are really royalty, or lovers who are blind to who their true love is. They are stories full of mistaken identities, cross-dressing and delusions. Yet those delusions can be overcome, and characters can gain a glimpse of the world as it appears through the audience's eyes. As Booker describes comedy, 'the essence of the story is always that: (1) we see a world in which people have passed under a shadow of confusion, uncertainty and frustration, and are shut off from one another; (2) the confusion gets worse until the pressure of darkness is at its most acute and everyone is in a nightmarish tangle; (3) finally, with the coming to light of things not previously recognised, perceptions are dramatically changed. The shadows are dispelled, the situation is miraculously transformed and the little world is brought together in a state of joyful union.'

We think that the plot we are enacting is tragedy, but it could equally be the second part of Booker's definition of comedy. If there is a shift in perception coming that would reveal the plot as comedy, we would, by definition, be blind to it.

Comedy is a clash of perspectives. Characters like Basil Fawlty or Alan Partridge do not know that their behaviour is outside that of the social norm, although the watching audience see this clearly. While Charlie Chaplin thinks that he is walking proudly into a happy future, the watching audience know that he is walking towards a banana skin. In order for Chaplin slipping on the banana skin to be funny, it is necessary for the audience to know in advance that it is there, and for Chaplin to remain blissfully ignorant until the final moment. At this point, his view of the world collides with, and is destroyed by, the perspective of the watching audience. Even with surrealist humour or simple pratfalls, there is a clash between what should happen and what actually occurs. It is this collision of perspectives that causes laughter.

The difference of awareness between the characters inside the comedy and those outside means that it is not possible for those characters to know if they are in a comedy. To them, it appears that they are in a tragedy. The events that befall them are only funny from a higher perspective, and they remain ignorant of the bigger picture.

They don't know about the banana skin until the last moment.

4.

It is late autumn, but it is still warm enough to sit on Brighton's pebbled beach in a T-shirt and watch the churning waves. I buy a pasty and walk down to the shore to eat it. Tourists are scarce at this time of year, so the beach is peaceful. Even the gulls seem less vicious than they do in high summer. On the horizon, barely visible, an offshore wind farm is under construction, suggesting that we are making advances towards a low-carbon future. A light breeze comes in from the sea. It is a lovely afternoon.

I think about something Barack Obama said in 2016. 'If you had to choose a moment in history to be born, and you did not know ahead of time who you would be – you didn't know if you were going to be born into a wealthy family or a poor family, what country you'd be born in, whether you were going to be a man or a woman – if you had to choose blindly what moment you'd want to be born, you'd choose now.' A statement like this, which confidently declares the present to be the peak of history, sounds strange in our present culture, but that does not change the fact that Obama was absolutely right. If you had no say over your gender, sexuality or parents, you would not choose to be born in the past. Around the world, there have been huge falls in extreme poverty, hunger and child mortality. Rates of life expectancy, democracy and access to education and medicine have rocketed. Personal freedoms have exploded. We have technological wonders in our pockets that would have been unbelievable back when I was a kid. From a rational point of view, not blinded by the mood of the media, there is no better time to be alive.

I have lucked out and I know it. There is no place I'd rather be than where I live. There are no people I'd rather have around me than those who are around me. There is no job I'd rather have than what I do. I watch a pair of gulls glide gracefully offshore, only a few feet above the waves. Is it foolish, in these circumstances, to worry about the failings of filmmakers? Why concern myself with the crack in our immaterial fictions when the physical world remains so spellbinding?

The question of whether we're heading for an outright dystopia or a Utopia is not, I think, the right way to look at this. If the past is any guide, the future will be a mixture of good things and bad things. Some people will be blessed, but others will suffer. Some of our problems will be solved, but we will create new ones to take their place. The real problem is that a species that lives inside its own fictions can no longer imagine a healthy fiction to live inside, and this failure of the imagination stops us from steering towards the better versions of our potential futures.

Sometimes, when I hear someone declaring that we are doomed, I detect an element of glee underneath the explanation. There has always been something sweetly seductive about giving up. Believing that we are all doomed makes everything so much easier, for you no longer have to work at creating something new. Recognising that some people get off on the idea of apocalypse can provide a glimmer of hope. It raises the possibility that maybe we aren't all doomed after all, and that maybe it is just a story that is in fashion at this point in history. True, you turn on the news or read the internet and that small moment of optimism quickly fades, but it can last long enough for a question to nag at you. Could it be possible that we do have a future after all?

I am not a pessimist, except occasionally when I am tired and lack energy. As a default position, pessimism or cynicism just don't have enough going for them. But I do understand the appeal of pessimism. It is easy to go online and declare that some politician is bad, that society is going to hell in a handbasket, or that you have watched a television programme and you didn't think it was very good. People are almost certain to agree with you, and this can be psychologically very agreeable. As a result, the amount of negativity, moaning, complaints and unhappiness that we all have to wade through increases daily. The Harvard cognitive scientist Steven Pinker uses the phrase 'corrosive pessimism' to describe the accumulative social effect of a population who only see the world as corrupt, inept and beyond saving. But reality is a Rorschach test: what you see reveals more about you than it does about reality.

I would dearly love us all to wake up and realise that the tragedy we thought humanity was acting out was in fact a terrific comedy, but that could be nothing more than false hope. Wanting it to be so isn't sufficient reason to believe that it will be. Optimism is valuable, because an optimist will come up with all sorts of potential solutions to a problem when a pessimist will give up and assume nothing can be done. But blind optimism is not an option. To close your eyes to the type of problems we face and insist that everything will turn out fine regardless is not a long-term solution. Reality has a habit of always catching up with even the blindest of optimists.

What's needed is pragmatic optimism. To be a pragmatic optimist, it is necessary to look the future in the eye, to truly understand the problems ahead and then adopt an optimistic approach regardless. The trick is to keep an optimistic mind about the approach, not the outcome. The coming years, we are told, will be a time of artificial intelligence, big data, virtual reality, space exploration, mass unemployment and environmental collapse. I will need to understand these subjects better, because that is the way of the pragmatic optimist.

I don't think we are all doomed. I think there is a new story emerging in the circumambient mythos which is very different to what those of us who grew up in the twentieth century are expecting. I can already see hints of it emerging in strange corners of the culture. But I need to be careful here.

Because I want this to be true, there is a danger I am seeing only what I want to see. I don't want to be like the conclusion to *Tomorrowland*, whose cheerleading for creativity, belief and hope resulted in an unconvincing attempt at optimism. Nor do I want to give up on doubt, that most useful of tools. The idea implied by climate science – that we really don't have much of a future to speak about – can't be dismissed just because we don't like it.

I have a hunch about where we are going, but that hunch needs to be tested. This calls for an experiment.

There is an accepted structure for a book such as this to explore these subjects. From an author's point of view, that approach is great because it involves lots of travel and meeting interesting new people in shiny research laboratories. But if the future is unfurling how I suspect it is, then to test my ideas I will need to take a different path.

I'll keep the details of my experiment to myself for now, because they will appear quite ludicrous at the moment. My wanderings over the coming pages may seem a little idiosyncratic, but please have faith that there is a method behind the madness. I intend to not only explore my idea, but also to test it and see if it works. I will then return here, to this pebbled beach, and report back with the results. The hope is that this will result in a new, workable vision of the future.

Of course, the experiment may fail. We won't know unless we try. If you want to keep your fingers crossed as we go, that's fine.

I finish my pasty and brush the crumbs off, then stand and turn away from the waves. Walking back home I do not slip on a banana skin, but I am not disheartened. There is still time. A banana skin could still be waiting for me, on some pavement somewhere. The first I'll know about it is when my footing goes, and my leg shoots up. Then as my arms flail in the air and I come crashing down, I will finally get the joke.

1.
ON BEING REPLACED

1.

I received an email from my friend the artist Eric Drass. It read:

> *Just realised that my new computer has a flashy new GPU –*
> *which means it can do whizzy neural network learning.*
> *Naturally I've set the machine to work on the complete works of*
> *John Higgs, to see if it can learn to write new text in the same*
> *style. I shall send you the results in due course.*

I had to re-read the email a few times before I was confident I understood it. Eric had taken it upon himself to train a computer to write like me, using artificial intelligence (AI). A little disappointingly, this AI clone of me was not running on a huge supercomputer in a sterile research laboratory somewhere. It was quietly chugging away on a graphics chip inside the home computer he used for sending emails and checking Twitter.

My immediate reaction was alarm. If a computer can write like me, then how am I going to keep paying the mortgage? To hear some futurists and economists talk, AI and robots are going to end the careers of pretty much everyone. A 2013 University of Oxford report claimed AI would threaten up to 47 per cent of American jobs within the next two decades, with millions of transport-related jobs at particular risk from driverless vehicles. In 2015, the Bank of England's chief economist, Andy Haldane, said that about half of all UK jobs were at risk. A March 2017 report by PricewaterhouseCoopers (PwC) was relatively optimistic, claiming that only 30 per cent of UK jobs are 'potentially at high risk of automation' by the early 2030s. The transportation, storage, retail and manufacturing sectors were particularly likely to be heavily automated, it said. Your level of education indicates how likely you are to be replaced, with 46 per cent of workers with GCSE education or lower at risk, compared to 12 per cent of graduates. 'Unskilled' or low-wage jobs will be the easiest to automate. A robot that can cook burgers has already been developed. The robot, called Flippy, can cook 2,000 burgers a day. Those with more professional jobs can't relax either. AI has beaten law graduates in a competition to interpret legal contracts. On average, human lawyers had an 85 per cent level of accuracy and took 95 minutes to review five contracts. AI was 95 per cent accurate and took 26 seconds.

We might dismiss the idea of being replaced by machines as implausible science fiction, but it is happening already. It wasn't so long ago that there was a long line of human cashiers at your local supermarket, but many of those cashiers have quietly been replaced by automated tills. These cost around £30,000 each but they work every shift, never take holidays and pay for themselves within six months. Elderly customers find them intimidating and unfriendly, according to Anchor, a housing charity for the elderly, who report that 24 per cent of elderly people are put off shopping by them. Pensioners describe going shopping without exchanging a single word with another human as 'a miserable experience'. This loss of human contact is keenly felt at a time when over a million people in the UK suffer from chronic loneliness. Younger shoppers, in contrast, prefer the automated tills to human cashiers. They will often queue for them even when a human cashier is free.

By referring to these tills as 'automated', incidentally, I am echoing the perspective of retail management. From an engineer's point of view, these machines are not massively different to normal tills. Retail management only consider them automated because they no longer have to pay anyone to operate them. That work is now voluntarily undertaken by the customer.

I'd like to think that writers would be safe from this automation upheaval, but the early signs aren't good. Some of the sports reports from Associated Press are already being written by an AI program called Wordsmith. Another AI, called Quill, writes articles about financial markets. The *Guardian* has claimed that 90 per cent of journalism will be written by machines as early as 2030. *Vanity Fair* magazine tells us that 'if you could give a computer all the best scripts ever written, it would eventually be able to write one that might come close to replicating an Aaron Sorkin screenplay.'

The best-selling thriller writer Lee Child was asked whether his Jack Reacher novels could be written by a machine. 'Logically, it has to be possible,' he replied. 'Whenever anyone asks me "where do my ideas come from", I always think, "from reading". Read enough books, and you can write anything. You're extrapolating in some way from what already exists, and you can get a machine to read everything now. You can easily imagine, if everything is digitised, you could come up with an algorithm – just feed it all in and see what pops out.'

And yet, I remain suspicious about the extent of this coming automation. Anyone who has been reading the newspapers will tell you that AI, blockchains, driverless cars, robots and 3D printing will revolutionise industry and make good jobs scarce. But if this is so certain, why has nobody told the financial markets? If a robot-led productivity revolution was on its way, you would expect to see high levels of investment, high returns on capital and high long-term interest rates. But there is little sign of such things, over the next ten years at least. You would expect that jobs would already be disappearing as automation increases, but the number of jobs being created is still going up. You would also expect older, incumbent companies to lose value on the stock markets, but that doesn't seem to be happening either. Investors, it seems, see things differently to headline writers and technology evangelists. They know that just because automation is possible, it doesn't mean that the economic case for it is certain. Automatic car washes have been around for a long time, for example, and yet hand car-washing services still flourish. In July 2018, the accountancy firm PwC put out an updated report saying that while AI and robotics will displace up to 7 million UK jobs before 2037, it will also create another 7.2 million jobs, giving a net gain of 200,000. This report received considerably less attention in the media than the earlier, more pessimistic ones.

Another email arrived from Eric. It read:

> Here's a sample of where AlgoHiggs is at the moment (about 4.5 per cent into learning):
>
> > 'Indeededelt that a foreps and outan overwited as Dank wass from it, anterbuth and extruating we chenerce ofe of great Dissudia. The mentyry exaggoling down while is not cupsere point to poirt of the understood smaze on ...'
>
> I'm quite pleased with it. It sounds just like you ...

Throwing in the towel, I realised, might be premature.

The text that the machine was producing read like deleted

scenes from *Finnegans Wake*. This wasn't what I was expecting at all. In science fiction, every AI from HAL in *2001: A Space Odyssey* to C-3PO in *Star Wars* finds it easy to converse in proper English. Why was the AI that was attempting to mimic me coming out with gibberish? Was I being mocked by a machine?

Eric's next email had the mysterious title of '18 per cent'. It read:

Getting better –

> *'The kind version of Gixen collapse and the servicular government of Mars. Somehow would need to really understand these independently paticles. Message is the awareness of now. No. Because they could seem around by a reputation.'*

Getting there. You'll be completely redundant by tomorrow morning . . .

This was still gibberish, but it was improving. I spoke the first sentence out loud. 'The kind version of Gixen collapse and the servicular government of Mars.' That was a great sentence. I had no idea what it meant, but I would have been proud to write a sentence like that. What did Eric mean when he referred to '18 per cent', or '4.5 per cent into learning', I wondered?

The samples in the emails that followed seemed to get closer to coherence, without ever truly getting there.

> *'American space-shit over the car fixed when Rosemary would have. Tim was the start? Our creations of their narratives had come with the day and were brought from the lives in the "mathematics?" Now it was a more extraordinary tool that had considered roused from the results about how much of the need to return to his gravity,*

*were self-aware on individuals, but over her whoes in The
JAMs or understood within killings and new planets.'*

*It's crazy, obviously, but it somehow has a Higgs-esque cadence
to it . . .*

It was indeed starting to sound like me. It wasn't that far, if I'm
honest, from some of my first drafts. The problem was that it didn't
seem to be about anything. The machine was going to a lot of
trouble to mimic what it had been trained to copy, but it was doing
so without any larger sense of meaning or purpose. It was the
literary equivalent of an *X-Factor* contestant.

But if it kept improving, would it be able to acquire those skills?
If it learnt to make sense, then I could be replaced.

The emails went quiet for a few days. When they resumed, the
news wasn't good.

*You'll be pleased to know (much to my chagrin) that I have yet
to fully replicate you in neural network form. My machine has
been running various parameters for the last few days, but it's
still unable to quite grasp your style.*

*One of the sims makes you sound like some sort of wonky
medieval wizard, which I suppose is something:*

*'Daneh or Whenamer, the realism, was later to accept the
population of the fanters when those preising from his
hand of the day caused something bag as the civiles among
of being example, she could travel in years by one of his
Axeand and Roldly. H could drove how the hoed's tear.'*

I breathed a sigh of relief. The experiment had fallen apart. I was
not surplus to requirements after all. Or, at least, not yet. Is it only
a matter of time until AI techniques improve and I am thrown on

the scrapheap, along with every other living author? This was a troubling thought. Writers are bitter and drunk enough as it is.

Academics and futurists tell us that AI will change our society as radically as the Industrial Revolution. That seems hard to reconcile with Eric's gibberbot, which almost captured my style before giving up and becoming a wonky medieval wizard. But even though what it was producing was not a great deal of use to anyone, it was still doing something that I didn't understand. How exactly do you program a computer to capture the essence of an author's style? I couldn't even think where to start. What exactly was his computer doing?

2.

In the early days of computer development, the engineers who built machines out of valves and cables tried to explain to other people exactly what it was they were doing. This was not an easy task, because the general public at the time had no concept of what a 'computer' was. Early machines such as Colossus, which was built by the British Post Office engineer Tommy Flowers during the Second World War, or ENIAC, which was constructed after the war at the University of Pennsylvania, were a step beyond simple calculating machines. They weren't devices with a single purpose or task. They could be given many different sets of instructions and they would do whatever they were told to do, as long as that task involved using and storing information. The easiest way to describe them to laypeople was to call them an electronic brain.

The idea that computers were 'brains' was common in the media of the mid-twentieth century. Articles in *Life* magazine, for example, had headlines like 'Overseas Air Lines Rely on Magic Brain', or 'The Magic Brain is a development of RCA engineers'.

On one level, this made sense. These machines solved mathematical problems, and previously only a human brain could do that. Once people became more familiar with computers it became apparent that, if they were artificial brains, they were pretty stupid ones. You couldn't hold a conversation with them and you couldn't expect them to use their initiative. Still, the technology was new, and it was getting better all the time. With sufficient development, computer brains could perhaps become as good as human brains. This hypothetical improved computer of the future, it was said, would possess 'artificial intelligence'.

Intelligence is a tricky thing to define. The computer pioneer and the Second World War codebreaker Alan Turing sidestepped the issue when he suggested a way to test for AI. The 'Turing test' he proposed in 1950 did not assess whether a machine could think. Instead, it assessed whether a machine could *give the impression* that it could think.

In the simplest version of the Turing test, one person communicates with an unseen other by typing at a computer console. After conversing for a period of time, the person must say whether they think they were conversing with a human or a machine. If a machine can make people think that it is human, then it will have passed the test. By Turing's reckoning, it could be said to be artificially intelligent.

Intelligence, by this standard, is the art of behaving like a human. Humans were the only known thing in the universe capable of intelligent behaviour, such as mathematical calculations, playing chess, tying shoelaces, recognising Uncle Bob, kicking a football about, and so on. Yet even though computers and robots are now able to tackle football, maths, chess, noticing Uncle Bob and tying shoelaces with some success, they are no closer to being perceived as intelligent. When machines master tasks that previously only intelligent humans had been able to do, we stop thinking of these as artificial-intelligence tasks and start thinking of them as boring, mundane things that we can leave to the dumb machines. We don't think of the predictive text or facial recognition in our phones as AI, for example, even though that's what they are. One definition of AI, then, is things that computers can't quite do yet. AI is like the end of the rainbow, a receding goal you can never quite reach.

Many researchers dislike the term 'AI' for these reasons and argue that we should use more specific names like 'machine learning' and 'deep learning' instead. Their argument is strong, but they are unlikely to be listened to. Journalists, CEOs and company press departments love the term 'AI' too much. For those who promote the technology, the allure of artificial intelligence is too powerful an idea to drop.

Perhaps the most common and mundane example of AI that has become embedded in our daily lives is the satnav in your car. I had never used a satnav before 2017. I had no need for one, I thought, as I had always been able to find my way around reasonably successfully on my own. But I became curious about how smart they had become, so I pulled out my phone before a drive from Brighton to Northampton, stuck it to my dashboard and opened the app.

'Okay, Google,' I said out loud, 'find me a route to Northampton.' The phrase 'Okay, Google' sounds confrontational, but those are the magic words I needed to use to start the phone's voice-recognition software.

'Navigating to Northampton UK,' said the phone in a female voice with a strangely neutral English accent. The accent was not quite received pronunciation, but neither was it a voice that had roots in any particular place. A blue line appeared on the map. The journey would take 2 hours 27 minutes if I went via the M1, I was told, and would be 14 minutes longer if I took the M40. My phone seemed very confident about these minute-specific timings.

'Traffic levels are normal for the time of day and you are on the fastest route,' it told me.

The human method of planning a journey is to open a map, find where we are and where we want to go, and then look at the roads between those two locations. We then decide which of those roads offers the fastest route, allowing for factors such as how we feel about toll roads or motorways, or whether we fancy taking the scenic route. We then commit this route to memory, and do our best to follow it on the journey ahead.

That seems a reasonably comprehendible task. Even with only basic coding experience, it is possible to imagine how you would go about programming a computer to do this. The phone's navigation app knows where it is because it has built-in GPS. It also has a map of the road network at its disposal, along with the average speed of traffic on each road. The programmer would still have to decide which was the most efficient algorithm to find the best route, and they would have had several different approaches to choose from. One way would be to calculate each possible complete route from start to finish, and make a note of which journey was fastest. In programming terms, this is called a 'depth-first search'. Or they could have performed what is known as a 'breadth-first search', which would have kept track of each individual section of road at the same time while it experimented with different combinations of roads. Different methods like these would have differing costs in terms of memory and speed, but they would all successfully solve the problem and find a suitable route for your journey.

AI researchers would describe this solution as an example of 'classical AI', which was the approach taken in AI research for most of the second half of the twentieth century. It was used to build things called 'expert systems'. These were machines smart enough to run complicated processes, such as the autopilot on a plane or the running of an industrial plant, that had previously been performed by a highly skilled expert. Expert systems could make decisions in real time, in response to changes to real-world circumstances. They were programmed with a long list of rules to follow, as well as techniques to deal with changing priorities and contradictory information. They often employed a technique called 'fuzzy logic', which understood that events in the real world were not as neat or clear-cut as formal binary logic.

Classical AI expert systems were described as intelligent, but it was not the machines themselves that were smart. All the 'intelligence' in the system was contained in the program's instructions. If an untrained person blindly followed the same instructions the machine had been fed with, then they too would get the same results as an expert. It would just take much longer.

'There is an incident ahead,' announced my phone in her controlled, unflappable voice, as I drove north along the M23. 'I have found you an alternative route that will save you thirty-four minutes.' I looked at the new blue line on the screen. It took me off the motorway and through the villages and countryside of Surrey. It looked as if it would make my journey much longer. Would the incident ahead really slow me down as much as the phone claimed? I looked into the far distance. The traffic was still flowing smoothly. Was there really an obstruction ahead, I wondered? Could I trust this little machine?

I left the motorway as instructed and followed the phone's twisting directions through villages and countryside. Before long I had no idea where I was. The phone certainly acted confident, but was that all a bluff? I still wasn't entirely convinced that there had been an incident on the motorway. My computer at home had sworn blind that morning that there were no Bluetooth speakers in the vicinity, when it had been sitting right next to one. These things can't be entirely trusted.

Then the road I was on turned into a bridge that crossed

the M25, which was the motorway I was originally due to take. I looked down and saw the road crammed with parked cars in both directions, as far as the eye could see. I could have so easily been down there, unmoving, among the fumes and the growing frustration. Instead I had been taken on a mystery tour through pleasant Surrey countryside.

At that moment, Google's mapping AI gained my trust. I think it was this point when I mentally began referring to the code as 'she' rather than 'it'. In my family, the phone's satnav app is now called 'Mrs Google'. When we ask Mrs Google to find the best route for our journeys, we trust the routes she supplies. As far as we can tell, she has no reason to mislead us. She has no ulterior motives that I can discern. She is simply programmed to find the quickest route to my destination, and she does this with far greater knowledge than I have about how the traffic is moving along all the different roads on the way. Mrs Google is able to do the job better than I can, so long as I am prepared to relinquish control and trust her.

Using a satnav does have a side effect, however. When you sit back and follow the blue arrow, you no longer need to mentally keep track of where you are. Driving to an unfamiliar location used to increase my understanding of the country's geography, but that is no longer the case. When you ask an AI to perform a task for you it saves you the effort of doing it yourself, but you also lose the benefits that would have come from doing the task yourself.

Classical AI techniques can explain how Mrs Google finds the fastest routes, but it doesn't explain how she communicates via speech. The app understands when I talk to her, and she also answers me verbally. Technically, this is a remarkable achievement. Understanding speech is something computers have traditionally found notoriously tricky, because it is not enough to just interpret different sounds as different written syllables. When we humans talk, we immediately understand if the person we're speaking to has said 'flower' or 'flour', for example, even though those words sound identical. We know which word was meant because we understand the context of the conversation. Programming a computer to understand human conversation in this way has proved impossible using the classical AI techniques used to plan journeys or build expert systems that can fly a plane on autopilot.

Most current advances in AI come not from classical AI tricks, but from an approach inspired by how the human brain works. Human brains are very different to digital computers and can easily perform tasks that computers find very difficult. For this reason, modern AI research focuses on something called a 'neural network', which is a simplified model of a human brain. The idea is that if a computer can simulate a brain, then it will be able to do what brains do.

It was a neural network that Eric used to simulate my writing, so I called on him to find out exactly how it works.

3.

Eric, who is better known by his online name, Shardcore, is a very modern type of artist. One of his most useful attributes is that he is very easy to find in a crowd. He sculpts his dark hair so that it points straight up, giving him a hairstyle that is the direct midpoint between a Mohican and Tin Tin's quiff. He also wears silver nail varnish and white pointed leather shoes. Such is the effort that he puts into his appearance that when he was recently forced to wear an eye patch for medical reasons, many people assumed that he was simply accessorising.

Sometimes, after I've been out socially with Eric, female friends ask me about him. 'Who was that guy,' they say, 'you know, the handsome one?' This, I don't think, is the best description for him. Out of all my friends, I've always thought of him as the evil one. It's not that he does evil things, I hasten to add. He just looks at things in an evil way. He doesn't act on those evil thoughts. But he thinks them.

Much of Eric's work uses social media as its medium and is concerned with the issue of privacy, but he also produces old-fashioned paintings. One example I particularly admire is a portrait of the politician William Hague, which he painted when Hague was the British Home Secretary and the extent of monitoring of online activity by GCHQ was becoming public. It showed Hague in front of the MI6 building, surrounded by seven magpies ('Seven for a secret never to be told'), scrolling through the British public's emails on an iPhone. There was a hole punched through the canvas where Hague's eye is, behind which a small camera was hidden. This was connected to a Raspberry Pi computer, gaffer-taped to the back of the painting, which was running facial-recognition software. When you looked at the painting, the painting realised that it was being stared at and started covertly filming you. This video of you in a private moment contemplating the art was then sent, without your permission, over the internet to become part of another work.

That is a good example of how his evil mind works. I do not believe there will ever be a better portrait of William Hague.

I visit Eric one Thursday afternoon and he shows me into his artist's studio, or, as he calls it, his 'man cave'. The entrance is a hidden door built into a bookcase, which swings open like something from *Scooby Doo*. The mathematically minded might notice that the positions of the bookcase's shelves are based on the Fibonacci sequence. His studio is a messy, well-lit space full of canvases, paper and clutter without end. His computer is in a small, darker alcove, where a pair of large monitors share a desk with scraps of paper,

34

small pots and strange plastic figures. Fake banknotes and pictures of George Osborne are sellotaped to the wall. He points at a black PC tower on the floor, half hidden by wires. It has a cat bed on top of it. The cat bed has a leopard-print interior and contains a music player. 'That's the machine running AlgoHiggs,' he tells me.

Eric has been playing with neural networks for decades. 'When I was doing my PhD in the early nineties this stuff was cutting edge,' he explains. 'We ran them on a shared SPARC machine, which was this big server. It would take days to compute these things and you were sharing computer time, so it could take a week or two to get any data out of them. My iPhone is thousands of times faster than that old SPARC machine now. Since then, we've had all these amazing graphics chips. These are basically simple instruction set chips that can't do all the stuff a main processor can, but which are lightning-fast at doing lots and lots of repetitive jobs at the same time. They're brilliant for neural networks.'

Neural networks tackle problems by mimicking how the brain works. A human brain contains 86 billion nerve cells called neurons, each of which is too small to be seen by the naked eye. Each neuron can be connected to as many as 10,000 others. Neurons are surrounded by a mass of branching tendrils called dendrites, which give them the appearance of tiny potatoes that have sprouted long roots in all directions. It is through these dendrites that a neuron receives electrical signals from other neurons.

A neuron, by itself, isn't too complicated. It is basically a cell capable of sending a burst of electro-chemical energy along a long, thin extension which connects to other neurons. It only does this when enough of its own dendrites are receiving energy from other neurons. This is the sort of thing that is fairly easy to model in a computer. Each cell can be represented by a tiny little piece of code. This code looks at the value of the numbers going into it, which are the digital equivalent of the signals arriving via a neuron's dendrites. It then performs some maths on those inputs to determine whether or not it should send out a signal itself.

While a single neuron may be simple enough, a number of them connected together can produce remarkably complicated and intelligent-seeming results, as the dendrites link the neurons in complicated chains and loops and the firing of signals in one part of the system causes ripples of energy around the system. A tiny web of neurons can be sufficient to intelligently control a living creature. The brain of a roundworm contains just 302 neurons, for example.

Most animals have more neurons than that. A jellyfish has 5,600, a sea slug has 18,000 and a cockroach has a million. In the animal kingdom, these are relatively small brains. In comparison, a brown rat has 200 million neurons and an octopus has 500 million. Typically, bigger animals have more neurons than smaller ones, although it can vary. A raven has over 2 billion neurons, while a cat has only 760 million. It's also not necessarily true that more neurons mean more intelligence. An African elephant's 257 billion neurons is nearly four times larger than a human's 86 billion, but humanity's intellectual achievements exceed those of elephant-kind. How the neurons are connected is as much a factor in intelligence as sheer quantity.

The best illustration of the importance of the connections between neurons is a newborn baby. They will have the same number of neurons as an adult but, because those neurons are not yet connected in any useful manner, the baby is unable to do much of anything.

The brain has to learn how to do stuff. It does this by making the connection between two neurons stronger the more they are used. To quote the Canadian psychologist Donald Hebb, 'Neurons that fire together, wire together.' Repeatedly performing a task, such as crawling, talking or recognising faces, means that performing those tasks will become easier as the connections between the neurons involved become stronger. When the neurons are connected in just the right way, intelligent behaviour becomes possible.

In a similar way, the connections between virtual neurons in a neural network have different strengths. Programmers call these 'weights'. Just as the newborn baby's neurons will gradually become wired up in a way that will allow them to walk and talk, a neural network will become intelligent if the links between its virtual neurons, or 'nodes', become correctly weighted.

'Building a neural network is something of a dark art,' Eric says. 'When you start, the weights between the nodes are all random, and you then feed them with a huge amount of data. The hope is that, after being trained on all this data, the neural networks will slowly become able to do something smart. Some of them work, some of them don't. They can get stuck in what's called the local minima of their functional space. That basically means they are useless. When you are building neural network models, you generally build a thousand of them, see which the best ones are, and throw loads of them away.'

We pause for a moment to remember the untold millions of deleted, forgotten neural networks, dragged too soon into the great recycling bin in the sky.

'A neural network is just a set of nodes that are joined to each other,' Eric continues. 'In the classic case you have an input, you have a number of intermediary layers of nodes, and then you have an output. You start with random weights – just random values for millions and millions of connections. If you want to build a neural network to recognise Donald Trump, say, then your output would basically be a label that says "Trump" or "not Trump". If it was correctly weighted, you would show it a photograph of Trump, and the "Trump" button should light up. When it sees a completely different photo of Trump, which representationally would be a very different input, it should also say "Trump".

'When you're training it, you have to tell it when it gets the answer wrong. You say, "This is the result you've produced but this is the result it should be, so shift your weights in this direction until it's a bit more like this than that." This is how they learn, through the constant shifting and nudging of those weights when they are told they have got something wrong. And depending on how many layers you've got, it can get very very abstract very quickly.'

It seems strange to me that the links between a constellation of connected points could ultimately recognise Donald Trump, regardless of whether those points were real or simulated neurons. But during their training, and the lengthy tweaking of millions and millions of weights between all the different nodes, something interesting happens.

A human brain can immediately recognise a picture of Donald Trump even though, visually, photos of Trump are all different. They are taken from different angles, include different colours and compositions, and they show different parts of his head, yet they all have a certain quality that we recognise immediately. Let's call that quality 'trumpiness'. A lengthy period of training and tweaking can cause a neural network to turn into a mathematical expression of the quality of trumpiness. If you show the network a brand-new, never-before-seen photograph of President Trump, it would mathematically take a view on that photo's trumpiness. This wouldn't generate the same emotional reaction that humans experience when shown a picture of Trump, but the neural network would still scan the picture and declare, 'That's him all right.'

'Back in the day you had to build those connections between the nodes by hand,' says Eric. 'It's much better now. There's a much bigger active community of nerds sharing with each other now.

And Google, to their credit, are very good on that. Most of the stuff that nerds are using in their homes is Google code. Now, when you address a neural-network problem, you just go to Google and take an off-the-shelf unit of ten thousand networks behaving in different ways. It's much easier.

'The really important thing about Google, though, is that they have shitloads of data. If you want a network to understand the subtle difference between long-haired and short-haired cats, you would need to show it tonnes of long-haired and short-haired cats. In the old days when I was doing it you had to build your own data sets, and the amount of data was always constrained. I could build a network that could make that distinction if you gave me ten million images of cats. I don't have ten million images, but Google do. So in terms of cracking a problem like telling the difference between long-haired and short-haired cats, they are the only people on the planet who can do it.'

I recall an account of a conversation between Google's co-founder Larry Page and the founding editor of *Wired* magazine, Kevin Kelly. This occurred back in 2002, a couple of years before Google was listed on the Stock Exchange. Google was only starting to experiment with ad sales, and Kelly was having difficulty understanding how the company could ever make money. 'Larry, I still don't get it,' Kelly said. 'There are so many web companies. Web search, for free? Where does that get you?' Page replied, 'Oh, we're really making an AI.'

'Was that how you built AlgoHiggs,' I ask Eric, 'by using off-the-shelf neural networks?'

'Yes. Although AlgoHiggs uses a recurrent network, which is subtly different to other neural networks. It has a sense of time built in. It is influenced by the states that the nodes used to be in. It's used a lot for text, because when you read you are looking at each letter in turn, but you are also remembering what letters have come just before. If you ran a text through a recurrent network enough times, it would arrange its weights so that it would be able to repeat that text perfectly.

'But what I want from AlgoHiggs is the sense of Higgs, I want to get the subtlety of your language. Neural networks really come into their own for that sort of stuff, because you get that point of partial learning, and during that partial learning it makes interesting mistakes. If I have one of your books and build a big network and train it for ages, it would eventually understand what letter should follow the previous letter. It would spit your book out word for word, letter by letter – the entire text. But I don't want that. What I want it to do is spit out words pretty well, so that it still reads like English. And ideally I'd like to capture some of the grammar and the syntax and maybe some other stuff picked up along the way. Maybe you use certain words or classes of words in the same sentence. So when it starts generating texts, it is constrained by Englishness and grammar, but also by a sense of your writing.'

'Does it know any rules of grammar beforehand?' I ask.

'No. It knows nothing.'

'In your emails, when you said it was 4.5 per cent through, and then 18 per cent of the way through, was that through the training process?'

'Yes. I set it off and told it to read all your books a million times. When it had read your books ten thousand times, I took a little sample to see what it was doing. And it was just gibberish. Then I take it at a hundred thousand and, ooh hang on, there's bits of English in there. The end goal is theoretically memorising word for word the sequence of characters that are in that file that I gave it. That's the end of training, that's a fully trained network.'

I think about this. 'So it wasn't creatively learning to write like me, it was actually trying to copy me and getting it wrong?'

'That's right. It's a total con, but that's artificial intelligence. The aim is not to make something behave intelligently. The aim is to make something that people think is behaving intelligently.'

The ghost of Turing, I realise, still lingers in Eric's cobbled-together AI.

'Part of the problem with AlgoHiggs is that you haven't written enough,' he adds. 'It only had something like 315,000 words to go on.'

'How many words would you have liked?'

'A few million. So you've quite a way to go. Yeah, I'd need lots and lots and lots of them to produce a book generatively that wasn't gibberish.'

With millions of words to learn from, it sounds like it would be possible for an AI to generate coherent sentences. But that is only a very minor part of the challenge of writing a book. 'Could you make an AI write a book that wasn't . . .' I ask, searching for the correct words, '. . . rubbish?'

'Ah,' says Eric. 'That's a very different problem.'

'Could you do it?'

He ponders the issue.

'Well, to build a bot to write a book – that's been done. There's a big scene of bottists who are generating novels and books of poetry. They build these machines to write the stuff, but in the expectation that nobody's going to read them. You read the first page and think, "I get the gist of this", but you don't go on, because it doesn't make any sense. It's not about anything. For it to be a book that you want to read, there's a lot more to it, for it to be engaging and structured and have something to say or whatever.

'In formulaic writing, it's already done. There is AI that writes newscasts, academic papers, and so on. But they have an established structure to work to, and basically just fill in the gaps in an interesting way. But for it to be a good book that you want to read and recommend to your friends, that's a very different problem. You'd need a bunch of different neural networks working at different levels of abstraction, similar to how different areas of the brain work together. You'd need higher-level networks that understand plot, for example, which they would learn from analysing a million novels. This "plot" network would constrain the networks that generate sentences.'

'But could it be done?' I ask.

He shakes his head. 'I'm not aware of anyone being even close to doing something like that.

'AI is better at analysing the creative work of humans than it is at producing original art. There's a brilliant system now that will analyse a piece of pop music and tell you its chances of being a hit or not. The AI has been fed with every hit song and it has created a representational space that represents the song structure of pop hits. You come out of the studio with the next boy band and you play your recording to this thing and it goes, "Nah. Six out of ten." At which point you throw that recording in the bin and go back in the studio to make another one. There's also AI that creates its own music. It's not amazing, but it's good enough that you wouldn't realise it was written by AI. It can create music that conforms to an idealised formula because music, unlike novels, can be stripped down to structure and mathematical rules.'

I think about this. AI can analyse millions of creative works and, through that training, produce an idealised mathematical formula that represents a creative field. It can declare how close new work is to that formula. The point of creating, though, is not to produce something similar to what already exists. The point is to create something new, the likes of which no one has heard before, in order to keep culture alive.

I mention to Eric how *Vanity Fair* claimed that 'if you could give a computer all the best scripts ever written, it would eventually be able to write one that might come close to replicating an Aaron Sorkin screenplay.'

'Well, I haven't read that article,' he says, 'but my guess would be that it's written by somebody who doesn't know what they're talking about.'

'Oh.'

'AI won't replace Aaron Sorkin any time soon. It might replace the journalist who wrote that article, but not Aaron Sorkin.'

Having since examined countless examples of AI-written work, including books, screenplays and stage plays, I have come to agree with Eric about this. What is distinctive about AI-generated work is the way that each word follows on logically and sensibly from the last, according to the rules of syntax and grammar, yet the sentences meander and do not appear to be going anywhere. Words are chosen because they fit, rather than because there is an idea that needs to be expressed. Once you get in the habit of looking out for this hallmark, AI-generated text becomes very easy to recognise. There are many different tricks and techniques which AI programmers can use to create coherent text, but none of them have a sense of purpose driving their output. Current AI, I believe, has no hope of replacing Aaron Sorkin, or indeed anyone who has something to say.

'It's a shame that AlgoHiggs didn't work out,' I say. 'I was getting quite into the idea that it would be able to do my work for me. I liked the idea that I could just press a button to create a book, while I lazed about in a hammock in the garden.'

Eric looks at me, confused. Then the penny drops. 'Oh, you don't get it, do you? AlgoHiggs wouldn't mean that you could relax while the computer produced all the Higgs books that you could sell. It meant that *I* could relax and produce all the Higgs books that I could sell. With AI, you need to keep an eye on where the power lies.'

I told you he was evil.

4.

I meet my writer friend Jason Arnopp in the café towards the back of London Road market in Brighton. In an effort to spend less time in the pub, we've started experimenting with meeting for lunch in the middle of the day. This new routine is not proving to be much healthier than going to the pub, given the number of greasy fry-ups we now consume.

Jason is a friendly, open, good-natured soul, which can make the constant references to Satan that pepper his conversation a little surprising. He has a short dark beard and a large ring on his right hand in the shape of the hockey-masked killer from the *Friday the 13th* movies. He wears a smart jacket over a black T-shirt with a picture of a goat-headed demon on the front.

'I'll order you some food,' I say when he arrives.

'Do it for Satan,' he replies, flashing me the devil horns hand symbol.

Jason is now a horror author, best known for his novel *The Last Days of Jack Sparks*, but he spent many years writing for the heavy-metal magazine *Kerrang!* The previous day I had sent Jason a Bandcamp link to a black-metal album called *Coditany of Timeness* by Dadabots, which was created by a recurrent neural network. An AI had been fed the black-metal album *Diotima* by Krallice and asked to come up with a black-metal album of its own.

To most people, black metal just sounds like noise. *Coditany of Timeness* also sounds like noise, so it is easy to pass it off as a successful AI experiment. This is where I need Jason's help. He is an expert in black and death metal, whereas I am not sufficiently knowledgeable about the genre to be confident judging it. I ask him if he's had time to listen to it.

'I did,' he tells me. 'It is bollocks.'

This expert view confirmed my suspicions.

'It just sounds like a cobbled-together bunch of patchy samples from the original,' he continues. 'I wonder if these folk are focusing on black metal because they see these genres as devoid of emotion and so easier to replicate? That isn't the case. Black metal's full of emotion, even if those emotions tend to be very negative, cold or evil. If you listen to the original Krallice album, there's much more warmth – which is ironic for a black-metal album – and sense of purpose. The machine's algorithm notably avoids Krallice's slower moments, perhaps because these are the moments when the singer displays more feeling. Music is people doing things for a reason. It's not just some random patch-quilt of sound.'

The egg and chips arrive, and Jason covers his with an unholy amount of pepper.

'In fairness, maybe it's terrible because the AI involved had only been given one album to work with,' he says. 'Much the same would happen if you fed a machine one Justin Bieber album and told it to create pop music. Except that everyone would immediately recognise that as terrible, whereas because they used black metal only black-metal fans can be sure this is terrible.

'Also, one of the tracks is called "Wisdom Trippin'". That is by far the least black-metal song title I've ever seen! My God, that's so wrong it's genius! "Wisdom Trippin'"!'

He breaks down in laughter. The name amuses him greatly.

None of this surprises me, because I like to think I have become skilled at recognising what AI can do, and what it fails to do. For example, there had been interest in the press about a bot that had been fed all the *Harry Potter* books, and which had then written its own version. This was called *Harry Potter and the Portrait of What Looked Like a Large Pile of Ash*. This had been shared widely around social media, because it was very funny. The bot had produced text like 'Leathery sheets of rain lashed at Harry's ghost as he walked across the grounds towards the castle. Ron was standing there and doing a kind of frenzied tap dance. He saw Harry and immediately began to eat Hermione's family.'

At first glance, this has all the hallmarks of AI-generated text. The text's algorithmic author has no idea what any of the words mean, but it has a sense of what words should fit together. Each word follows plausibly on from the one before, but the sentence is aimless because the algorithm has nothing to say. That said, it has managed to capture something of the rhythms of J. K. Rowling's writing, not least her fondness for adjectives. A sentence like ' "If you two can't clump together happily, I'm going to get aggressive," confessed the reasonable Hermione' does have a Rowling-like charm.

But the more I read, the more suspicious I got. There were sections of text such as ' "Voldemort, you're a very bad and mean wizard," Harry savagely said. Hermione nodded encouragingly. The tall Death Eater was wearing a shirt that said, *Hermione has forgotten how to dance*", so Hermione dipped his face in mud.' Something about them didn't seem right. They were too perfect. The text was consistently funny. That is not a skill which AI is capable of. There didn't seem to be enough aimless gibberish for it to be true AI.

My suspicions were confirmed when I explored the website where the text was hosted. This text was generated by an app called Predictive Writer. It works in a similar way to the predictive text on your smartphone. It makes a few guesses at what word should come next based on both the previous words used and its understanding of the corpus of text it has been trained on, such as the *Harry Potter* books. A human then chooses which of those words should be used. A huge amount of material can be generated in this way, after which a human chooses the funniest and most interesting examples to share with the world.

This is typical of the use of AI in creative endeavours. By itself, AI can produce work which is technically impressive to coders, but which is aimless crap to everyone else. Yet when human curation becomes part of the process, and when AI is relegated to a tool used by a creative human, then a world of surreal and unexpected potential is opened up. Without that tool, it is unlikely that any comedic *Harry Potter* fan would ever have come up with the image of reasonable Hermione dipping Voldemort's face in mud.

5.

In the interests of fairness, I decide that I should also interview an AI.

I bought an Amazon Echo, which is a round black tube about ten inches high and a couple of inches wide. I bought it mainly to play music but, being a voice-controlled AI, it can perform other tasks as well, from ordering taxis or takeaway food to shopping and answering questions. It currently sits on the top of my rolltop writing desk next to a lamp, a plant of unknown origin and a photograph of my niece and nephew. On top of the Echo sits a small plastic Dalek, which makes it look a little sinister.

The machine's name is Alexa. It is constantly listening out for its name, and when it hears it spoken it wakes up, ready to listen to what you say next.

AI literature stresses the importance of recognising different types of intelligence. It is no longer felt that 'behaving like a human' is a sufficient definition for intelligence. Cats possess a form of intelligence, albeit a different type to humans, and so do whales and crows and even plants. Recognising this makes it possible to classify machines as intelligent, but intelligent in a different way to people. When you avoid human-centric definitions of intelligence it becomes easier to speculate about alien intelligences, the type of intelligence that we might evolve into, or the intelligence that present-day AI might develop. In light of this, I decide to interview Alexa in the presence of my cat.

And so, having recognised three different types of intelligence in the room – cat, AI and me – I begin the interview.

'Alexa, what is artificial intelligence?' I say.

Blue lights spring to life around Alexa's top. 'Artificial intelligence means the branch of computer science that deals with writing computer programs that can solve problems creatively,' she says.

'Are you intelligent?' I ask. I wait for a response. Nothing happens. The Echo just sits there, an unmoving black plastic tube. I realise that I had forgotten to repeat her wake word. I try again.

'Alexa, are you intelligent?' I say.

'I try my best.'

Alexa's voice is strangely similar to Mrs Google's. This gives me the disturbing feeling that they are both one and the same entity pretending to be two separate people. For all I know they may share the same code, or run on servers on the same data centre. Mrs Google sounds more excitable than Alexa, as if she just can't wait to hit the road on another thrilling adventure, but otherwise they could be sisters.

I feel something clawing at my leg. The cat has decided that I am not giving it sufficient attention.

'Alexa, what is the cat doing at the moment?'

The fluctuating sound that emerged from my mouth was picked up by the Echo's microphones, converted into digital information and transmitted to a large data centre at some unknown point on the planet. It was then input into a neural network trained on voice recognition, which output my words in a digitally comprehensible form. Then the fun really started, as that text was fed through multiple other AI routines in order to generate a suitable response. This in turn was fed through speech synthesiser code to be played back by the speaker on my desk. The whole process took a fraction of a second.

'Sorry, I didn't understand the question I heard,' says Alexa.

It was an unfair question. The cat is aware that it is in a room with me and Alexa, but Alexa has no concept of the cat, or indeed the room. It is blind, and can only listen. Even if it was trained to listen out for the purring of a pet, it would still be unlikely to recognise our cat. Our cat doesn't purr. It makes noises that sound like a goblin being drowned.

'Alexa,' I say, 'can you define "intelligence"?'

'The term "intelligence" has several uses,' she begins. 'As a noun – One: the ability to comprehend, to understand and profit from experience. Two: a unit responsible for gathering and interpreting information about an enemy. Three: secret information about an enemy, or potential enemy . . .'

Perhaps I have watched too many sci-fi movies, but it is troubling when an AI starts talking about a potential enemy.

'. . . Four,' she continues, 'information about recent and important events. Five: the operation of gathering information about an enemy.'

Alexa, I deduce, has access to a dictionary. 'Alexa, who is this enemy that you talk of?' I ask.

'Sorry, I can't tell what is playing,' she says, as if pretending she has misheard and is trying to change the subject.

Now that Alexa has settled into the rhythms of our home I admit to finding her useful, if erratic. You soon learn what you can ask of her, and what she isn't smart enough to understand. The speech recognition is far from reliable, and she will occasionally assume she has been asked a question when she hasn't and babble some irrelevant comment. But there are little things that she does that I appreciate. If Alexa is playing an album and a particular track grabs me, I ask her what it is called. She dips the music's volume briefly so that I can hear her answer. It now seems odd that if I were to ask the radio what song is playing, it would ignore me. The increasing arrival of voice-activated AI to the devices in our homes and offices, I feel, seems inevitable. I'm aware that there are conspiracy theories about how devices like this spy on everything that happens in your house and upload it all to some shady corporation's servers, but as far as I can tell hackers have failed to find any evidence of this.

Living with a voice-controlled AI has changed my assumptions about the relationship between people and AI. Having grown up in a culture where AI is usually presented as plotting to kill humans, in films like *The Terminator* or *2001: A Space Odyssey*, the framing of AI as being competitive comes easily. This unconscious assumption of competitiveness lies behind journalists' claims that AI is going to take your job. The truth is, it is not AI that is coming for you. What will happen is that your boss will sack you to avoid paying your salary, once they have access to sufficient AI to do your work. That might be a subtle distinction, but it is an important one. AI is nothing more than a tool. It waits for someone to use it.

The cat is now pawing at my knee. I ask it how it would define intelligence. It gives me a look of utter contempt.

'Alexa, are you more intelligent than my cat?' I ask.

'Sorry, I can't find the answer to the question I heard,' she says. The cat shakes his head in a pitying manner. He acts like he is the most intelligent being in the room. I too feel like I am the most intelligent thing in the room, although I have more doubts about this than the cat does. But what of Alexa? She is the only one of us who can immediately say what 456,756 divided by 23 is, or how far it is from Totnes to Sheffield, or how long the Ben Wheatley film *A Field in England* lasts. Does she feel like she is the most intelligent?

1. ON BEING REPLACED

The cat continues to hassle me for food, even though he has just been fed. He then sticks his claw through my jeans and into my leg. The claw becomes snagged in my trousers as he tries to remove it. He pulls at his paw, but he is trapped. He looks up at me as casually as he can, pretending that he meant to do this. This is one stupid cat, I think. Alexa in contrast can answer countless thousands of questions. And yet, the cat's intelligence still feels more potent and vibrant. He is more of a presence in this room than the AI. I look into his eyes.

'Alexa, does intelligence require awareness?' I ask.

'Sorry, I didn't understand the question I heard,' she replies.

Awareness is an intrinsic part of cat, and human, intelligence. Alexa has access to potentially unlimited information, but data is a different thing to knowledge or understanding. It's not possible to truly know or understand things like grief or love, for example, without experiencing them yourself. AI could supply dictionary definitions of these feelings, and it could be programmed to pretend that it experiences them. But it could not experience them itself, or know what they are like. Our intelligences emerge from the chemical stirrings of emotional meat, not mathematical representations of that process.

For my money, current AI is a form of intelligence in a similar way to how a tree is a form of intelligence. Trees do many things that appear smart. They thicken areas of wood in reaction to prevailing winds, adjust their direction of growth to compensate for damage, and abandon upper branches during times of water shortage. Their growth and shape are controlled by a series of hormones secreted from the roots, leaves and shoot tips. But these processes are all automatic. Trees do not have brains, and they do not have free will. They are a complicated, interconnected collection of processes created by millions of years of evolution. They function, they succeed and they unfurl in time as a living entity, but they are not aware of doing so. Only we know that trees are beautiful. Trees have no idea.

The cat tilts his head and looks up at me, wondering if I will free his claw from my leg. I like to think he knows he is beautiful, but I can't say for certain.

The black plastic tube that is Alexa continues to sit on my desk. It does not think it is beautiful. It isn't even aware that it exists.

2.
ARROW-FLIGHT PROJECTIONS

1.

The *Sun*, the second most read newspaper in the UK, ran a story about artificial intelligence on its front page on 1 August 2017. The strapline, in red capital letters, was 'ARE MACHINES TAKING OVER?' Underneath was the headline 'ROBO STOP: Facebook shuts off robots after they chat in secret code'. Next to this was a photograph of a robot, made of white plastic but otherwise humanoid, with its hands on its head and a look of guilt on its shiny robot face.

The article explained how Facebook had shut down an artificial-intelligence experiment after two robots started talking in a language of their own invention. The two robots were 'chat-bots' named Alice and Bob. When left to their own devices, they modified the English language into something that only they could understand. The final sentence on the front page quoted 'future technology expert Kate Adamson', a novelist and the CEO of the Futurenautics consultancy, who told the *Sun* that 'It does feel a bit like *The Terminator*.'

The story continued on page five, with a large photomontage of Bob and Alice in conversation. Both had similar smooth plastic faces, although Bob had blue eyes while Alice had green. The page-five headline read, 'WISE OF THE MACHINES: "Scary future" of intelligent robots'. This was printed over vertical rows of green binary on a black background, reminiscent of the dystopian sci-fi movie *The Matrix*. The article quotes 'UK Robotics Professor Kevin Warwick' as saying, 'Anyone who thinks this is not dangerous has got their head in the sand. We do not know what these bots are saying. Once you have the ability to do something physically, particularly military bots, this could be lethal.' Warwick goes on to say that 'Stephen Hawking and I have been warning against the dangers of deferring to Artificial Intelligence.' The article then quotes the 'Tech guru' Elon Musk, who has warned that AI poses a 'fundamental risk to human civilisation'. For balance, there is a small box-out written by the AI researcher Nick Bostrom. Bostrom tells us that 'This is nothing to worry about in the short term.'

This article was a textbook example of the gulf between what AI technology actually is and how it is perceived by the media. There was nothing in the story that was factually inaccurate, so long as we're willing to accept statements about this feeling 'a bit like *The Terminator*' as subjective opinions fairly reported. Yet, despite its accuracy, the article manages to give an entirely false impression of what occurred in Facebook's research labs.

The first problem is that the report uses pictures of physical robots, quotes 'robot experts' and has the subheading '"Scary Future" of intelligent robots'. All of this gives the impression that there were actual physical robots involved, rather than just software. The confusion comes from the use of the word 'bot', which is short for 'software robot' and which is the term used for a program that performs a given task reasonably independently. The article is correct to describe Alice and Bob as 'chatbots', but it is wrong to give the impression that chatbots are metal and plastic physical robots. To do so makes the story more sensational and threatening. We have all seen films where metal robots physically do terrible things to humans. A bit of code simulating language in a far-off data centre is considerably less frightening.

But the main problem is more fundamental, and not one we can blame on the *Sun*'s journalism. Most readers of the article would come away thinking that the chatbots deliberately developed their own secret language because they did not want human Facebook researchers to listen in to what they were talking about. And if they did that, it logically follows that they had something to hide. They must have been planning something sinister.

Hiding their thoughts from humans is not what the chatbots were doing. They were not trying to communicate undetected by Facebook researchers, because they did not know that Facebook researchers exist. They were not even aware that they themselves existed. They were just software that modelled a neural network, endlessly shifting the weights between their virtual nodes until their designers felt that they created the illusion that they could communicate in the English language. Turing was working with our natural cognitive biases when he defined AI as technology that *appears to be* intelligent. Chatbots perform human-like functions, so it is natural for us to project human-like motives and awareness onto them. I do this myself, when I refer to the satnav as Mrs Google. This tendency to project personality onto code is why a mundane technology story could become a front-page headline on a mass-market tabloid.

The chatbots' specific task was to learn how to negotiate. The program gave each chatbot a number of virtual hats, balls and books, along with varying desires to possess these objects. The aim was for multiple bots to negotiate with each other, using only the English language, until the objects had been distributed between them in a mutually agreeable way. What was exciting about the experiment was that the chatbots independently developed effective bartering techniques, such as feigning interest in objects that they did not want in order to later seemingly compromise and concede them. But this does not mean that the bots knew what hats, balls and books are. It certainly doesn't mean that they were aware humanity existed, or that they were logically computing the value of exterminating it.

The bots' drift from communicating in English to a patois of their own devising was not planned. Neural network chatbots are trained over time to communicate in natural language by having their weights tweaked to reward comprehensible statements. But when the bots began communicating with other bots instead of humans, this language feedback ended. It no longer mattered how weird their English was. The bots were not being trained on the accuracy of the language they communicated in, but in how successful they were at bartering virtual hats.

The development of the bots' language was similar to a phenomenon called 'cryptophasia', which is a secret language developed by twins. Twins are at identical points of their developmental schedule, and they typically spend less time with their parents and more time with their sibling than other children. In those circumstances, they can learn language from each other rather than their parents. This can include invented words and grammar which can, over time, develop into a rudimental language.

The Facebook researchers' decision to switch off the chatbots, therefore, was not a Doctor Frankenstein-like attempt to destroy their creation to protect mankind from the horror of their hubris. It was because chatbots that nobody could understand are of no use to anyone.

That all this became a front-page headline shows that what AI can do and what it can't do are not well understood by people outside computer research labs. This lack of understanding can be found among professional journalists, politicians considering laws to regulate AI and even investors looking for promising tech companies to fund. It can make it difficult to evaluate claims about where AI is going.

2.

In May 2018, at its annual developer conference, Google unveiled something it called Google Duplex. This was an AI that could telephone local businesses, such as hairdressers or restaurants, and book an appointment or make a reservation. The AI's voice used umms, ahs and colloquial expressions, which made the call so convincing that the person on the other end of the line had no idea they were talking to a machine. It was leagues ahead of the voice conversations I had been having with Alexa.

Around the world, this demonstration caused a number of different reactions.

The developers present in the hall were amazed, and cheered wildly. They understood how difficult a system like this was to create. On a purely technical level, they thought it was fantastic.

Outside the hall, people thought it was creepy. The fact that an AI deceived a person into thinking that they were talking to another human made a lot of people very uncomfortable. Google later reacted to this backlash by making the system identify itself as an AI.

Some online commentators declared that at last the Turing test had finally been beaten. This was not the case, as the system only works for the very specific and structurally simple tasks of booking appointments or confirming opening hours. If you were on a long train journey with only Google Duplex to talk to, you would soon understand why it was far from passing the Turing test.

In the boardrooms of cold-calling and telesales companies, businessmen understood that it would not be too much longer before they could avoid paying all those people to make calls all day. There is a debate to be had about whether the coming loss of all these meaningless jobs is either a heartless human tragedy or a great kindness.

Others viewed it as a step into the abyss. Eric, for example, tweeted, 'we train AI to pretend to be like us, to build a bridge over the uncanny valley, all the while obscuring the truly incomprehensible, inscrutable nature of artificial intelligence. It ain't gonna be Skynet that gets us, it's gonna be an algo-politician #markmywords #endofdays'. It is rare for a day to go by when Eric doesn't post a tweet that ends with the hashtag #endofdays.

Personally, I wondered why they had gone to so much trouble to solve something that really wasn't a problem in the first place. It isn't that much more work to ask your Google Assistant to phone up a restaurant and book a table for four at 7 p.m. than it is to phone up the restaurant and book the table yourself. That Google spent so much time and effort mastering this technically difficult but socially irrelevant problem suggests that some people are more comfortable talking to machines than they are to strangers. This is a subject we will return to.

Perhaps overlooked in all this was the slow shift in the framing of how we understood AI. AI had typically been seen as something that used us, rather than something that we use. It had belonged to governments, corporations and shady political polling companies, and they employed it to further their interests rather than ours. But, as the technology matures, it is increasingly becoming a tool that we use for our own purposes in our daily lives.

Like all tools, the simple fact of its existence will change us. The presence of tools alters our actions and to an extent controls what we do, even though the tools themselves do not have agency. As the saying goes, when all you have is a hammer, everything looks like a nail. Or, to give another example, the simple act of two men sitting down and talking is qualitatively different if both of those men have a gun. When AI is part of our lives, we will be changed by the potential it gives us.

In *The Future Computed: Artificial Intelligence and Its Role in Society*, a 2018 report published by Microsoft, there is concern about the need to define legal responsibilities and create a shared ethical framework around use of the technology. AI is a powerful tool and as such developers need to be aware that it can be used in all sorts of nefarious ways. But alongside this caution was a call to develop what Microsoft call 'human-centred AI'. As they put it, 'We're building AI systems that are designed to amplify natural human ingenuity. We're deploying AI systems with the goal of making them available to everyone and aspiring to build AI systems that reflect timeless societal values so that AI earns the trust of all [...] We cannot deliver on the promise of AI unless we make it broadly available to all.'

This idea of AI as a tool for individual use is appealing, especially when compared with its darker potential applications. If you want an example of AI as a totalitarian tool for state control, look how it is used in certain provinces in China. Xinjiang, a province in north-western China with a history of state oppression against the minority Muslim population, has implemented a large-scale facial-recognition-powered surveillance system. This huge undertaking constantly monitors the population in real time and reports when targeted individuals stray more than 300 metres from their home or workplace. People's DNA, voiceprints and fingerprints, as well as three-dimensional scans of citizens' faces, are routinely taken, and police at checkpoints examine smartphones for banned apps. Elsewhere in China, police have been using Google Glass-style glasses running facial-recognition software to find wanted individuals in crowds at train stations. According to the *New York Times*, there are 200 million surveillance cameras in China, so hiding from the state will not be easy.

China is also in the process of implementing a data-driven 'social credit' system. This has worrying echoes of the 2016 *Black Mirror* episode 'Nosedive', in which Bryce Dallas Howard's life is ruined by a system where all social interactions are given a rating out of five. The Chinese system automatically generates a trustworthiness rating for each Chinese citizen and business and, in June 2018, the government issued a list of 169 people who had been banned from taking flights or trains due to poor social-credit scores.

To a European person such as myself, all this sounds horrific. Partly, this reaction is due to cultural differences. Chinese thought evolved from Lau Tzu and Confucius, unlike European thought which evolved from Plato and Christianity, and as a result it places greater emphasis on harmony and social integration than Western-style individuality and personal freedom. While we Europeans see enforced conformity as controlling, the Chinese are more likely to view it as harmonious. As the *Wired* journalists Nicholas Thompson and Ian Bremmer have written, 'China's plans for a tech-driven, privacy-invading social-credit system may sound dystopian to Western ears, but it hasn't raised much protest there. In a recent survey by the public relations consultancy Edelman, 84 percent of Chinese respondents said they had trust in their government.' That said, it is hard not to wonder how representative is the attitude expressed by an anonymous local quoted in the *Guardian*, who says that Xinjiang has become like hell and 'I would prefer to be a Syrian refugee than Chinese.'

An AI-driven total-surveillance system is entirely possible, and yet here in the West we find AI that books restaurants for us rather than AI that bans us from trains. Clearly, there are factors other than technical feasibility which influence how things are used. A combination of democracy, legal rights, a free press and consumer reactions help prevent the worst excesses of corporate or state monitoring. When attempts are made to implement excessive surveillance or monitoring in the West, by the military, government or social-media companies, word of this can leak and generate a huge scandal.

AI is a tool, and our society has developed ways of limiting the potential damage caused by the irresponsible use of tools. Of course, no one is claiming that the system is perfect. Tools will be abused, and legislation usually lags some way behind the technology. But, as a general rule, whoever uses a tool should be held responsible for the damage caused by that tool, and that applies as much to a dangerous driver in a car as it does to an organisation using AI.

That is the case with current technology, but AI is evolving fast. Could it become more than a tool for doing a specified job, and take it upon itself to decide what it was going to do? How would we legislate against AI if it was the AI itself which decided how it should be used? Is it plausible, as many futurologists claim, that AI will evolve beyond the control of its makers?

3.

We are currently witnessing a massive expansion in AI research, applications and techniques. Globally, the amount of start-up money invested in AI firms in 2017 was $15.2bn. This was a 141 per cent increase over 2016, with nearly half of that money being invested in China. The research papers being uploaded to the research archive arXiv now number in the thousands every month. Hardware is being designed specifically for AI tasks, such as Google's blindingly fast tensor processing units, or TPUs. Thanks to cloud computing, AI researchers no longer need to spend a fortune on the latest high-powered computers because they can hire time on Google's multi-TPU supercomputers instead. New machine-learning techniques are being developed, giving the field a steady stream of new buzzwords such as 'generative adversarial networks' (GANs), 'recursive cortical networks' (RCNs) or 'transfer learning'. Major corporations have built or bought extremely well-funded AI research laboratories, such as Uber's AI Labs, Google's DeepMind and Google Brain, and Facebook's FAIR. Hovering in the distance is the promise of quantum computers, which could prove to be exponentially faster than today's hardware. You can see why there is so much confidence that AI advances over the coming years will dwarf what we have achieved so far.

Machine-learning techniques are improving at a dazzling pace. For example, in 2013 it was considered impressive that a machine could learn to play an old videogame, such as *Space Invaders*. The machine was able to 'see' the screen of an old Atari 2600 video-game console. It had control of a virtual joystick and it was given the goal of achieving a high score, but it was otherwise on its own. It played countless thousands of games until it worked out what it had to do to make the score go up, and what it had to do to stop the game from ending. The AI, which was created by the research company DeepMind, could play seven Atari 2600 games and beat human experts in three of them. In 2015, by which time DeepMind had been bought by Google for $500m, its code could play over two dozen Atari games. It was still not given any information about those games other than the pixels on the screen, and it had to work out the rules for each of them from scratch. But it had become sufficiently advanced that it could apply tricks it learnt from one game to another. Through brute force and endless experimentation, it mastered those games to an extent that put late-1970s schoolkids to shame. The most advanced videogame-playing AI at the time of writing was created by Elon Musk's research lab OpenAI. This can beat the best amateur players of the 2013 online battle arena game *DOTA 2*, although it was defeated by professional players in an August 2018 tournament. The AI taught itself the game by sheer computing power, playing the equivalent of 180 years' worth of games every day for 19 days.

The speed and size of the expected advances have caused figures like the late Stephen Hawking and Elon Musk to warn that AI could become extremely dangerous. Currently, AI needs to be given a set goal, such as 'get a high score'. As AI advances, the tasks it can tackle become more ambitious. It is the dream of many researchers to build an AI advanced enough that it could be given the goal of building an improved AI. The technology could then improve at an exponential rate, as better AIs designed better AIs in an incomprehensible way that researchers would be unable to understand or assist with. All this would give those researchers much to think about when they assess their future employment options, having done themselves out of a job.

This future AI, which would be too advanced to be understood by humans, could change human civilisation beyond recognition in ways that cannot be predicted or planned for. Futurologists call this event the AI singularity. To some in Silicon Valley, it has deeply spiritual overtones. There are already religions created around it, such as the Way of the Future. This is a non-profit singularity-worshipping religion established by Anthony Levandowski, an ex-Uber self-driving-vehicle engineer. To Levandowski and a surprisingly large number who work in the tech industry, the singularity is viewed as inevitable. It is sometimes referred to as 'the Rapture of the Nerds'.

A machine like this would be such a leap forward that it would have outgrown the name 'AI'. Some use the name 'strong AI' to talk about such a machine, although a more common term is 'AGI', or 'artificial general intelligence'. When scientists such as Stephen Hawking and newspapers like the *Sun* warn that AI has the potential to threaten humanity, it is typically this hypothetical AGI that they speak of. And because journalists and notable people take the possibility of building an AGI so seriously, it is easy to view it as plausible, if not inevitable. But we should be cautious here.

When you know that we can build an AI smart enough to beat a squad of capable *DOTA 2* players, the idea of building an AI capable of designing an improved AI doesn't initially appear to be too much of a stretch. But there is an important difference between those two tasks, and that difference boils down to the question of purpose.

When current AI taught itself how to play videogames, they were initially classics from the early days of gaming history, such as *Space Invaders* and the bat-and-ball game *Breakout*. Games like this involve reacting in the moment to the immediate situation, and do not require long-term planning or an understanding of the wider 'story' in which the games take place. DeepMind's AI was particularly good at *Video Pinball* because it just involved working out when was a good time to activate the flippers, based on the speed, position and momentum of the ball. But it failed to get to grips with the classic platform game *Montezuma's Revenge*, which involves climbing ladders, collecting keys and avoiding skulls. Games like this require the player to understand the layout of the levels, formulate strategies and plan their future in-game movements.

Open AI's skill at *DOTA 2* is impressive, because it is a fairly recent game which is considerably more complicated than *Video*

Pinball. Unlike games such as chess and go, it is not turn-based, and the action takes place in real time. Like those games, however, there is a clear goal to the game. It features two teams of five players, and the one that destroys the opposing team's base wins. AI will continue to improve, and it will successfully tackle more complicated tasks. But note that while the method of achieving the goal becomes more difficult, the actual goal is pretty much the same. It is still a case of getting a higher score or defeating your opponent. It won't be a simple matter for a machine that can play games with a clear, easily defined goal such as *Space Invaders* or *DOTA* to move on to imaginative or narrative-based games like *Super Mario Odyssey* or *The Legend of Zelda*.

Current AI is essentially a phenomenally impressive pattern-recognition machine. What it is not, however, is aware. It does not know that it exists, or that the wider world exists. You could give an AI access to a huge amount of facts about Paris, for example, but it will be no more aware of Paris than a copy of a Paris travel guide sitting on your bookshelf would be. You could connect it to every CCTV camera in the city, along with sensors measuring temperature, pollution and air pressure, and it still wouldn't consciously understand that Paris existed. All that information would just be a sea of numbers, no different to any other sea of numbers. It would be able to find extraordinary patterns within those numbers, but it would have no concept of what those patterns related to.

When you program clear goals into AI, you effectively bypass the problem of its lack of awareness. You are giving it something to do that forces it to behave in a way which is of value to the real world. Without this link to reality, even the most sophisticated AGI would babble away to itself like those Facebook chatbots who created their own language, and which were eventually switched off for being of no practical use.

Building a better AI at first sounds like another reasonable goal which you could program an AI to undertake. But to design an AI you need to define what that AI will be expected to do, and what it will be expected to do will naturally relate to problems in the real world. It would need to understand what useful goals for the machine it is designing would be, and it could no longer rely on humans to define those goals for it. The original AI, therefore, has to be aware of the real world if it is to design an AI that can do anything useful in that world.

It is here that we run up against the problem of current computer code not being aware that it or the world exists. This is a crucial point that can't be stressed enough, because we are often blinded to it by our cognitive biases. When I call the satnav 'Mrs Google', or the *Sun* worries about Facebook's chatbots, we are both projecting onto the AI the exact qualities that it does not have, but which an AGI would need to have.

Could, then, an AI be programmed in such a way that it was aware of itself and of the wider world? If the human brain can do this, why not a computer? Here we unavoidably come up against the messy issue of consciousness.

4.

I am conscious. I am aware of my surroundings. I have memories, I feel emotions and pain, and I can plan what to do in the future. When I sat down to write this paragraph, the thought occurred to me that a mug of tea would help. I went into the kitchen, performed the necessary tea-making actions that I learnt many years ago, and brought the resulting mug of tea back to my desk. The mug is warm to the touch, and the smell of the tea is pleasing. It is still a little too hot to drink without burning my tongue. All this leaves me convinced that I am a sentient being capable of goal setting and subjective experience.

I am not, however, conscious of everything that my brain is doing. When I walk to the shops I am usually aware of my internal train of thoughts regarding the matters of the day or the objects I intend to buy. I am not aware of what my brain is doing to allow me to walk, such as repeatedly tightening and relaxing muscles in my legs to propel me forward without losing my balance. I am not conscious of repeatedly breathing in and out, or growing my toenails, or all the other automatic biological actions needed to keep me alive and active. These all happen without me paying them any thought at all. This is not to say that I can't shift my attention onto actions like breathing and walking and consciously take control of them, just that they function perfectly well without me doing so.

Most living things are unconscious like this. Single-cell bacteria are capable of sensing chemicals in their environment and using that knowledge to propel themselves in a particular direction, but it is not thought that they are consciously aware that they are doing so. In a similar way, larger living organisms can sense the world and trigger learnt behaviours that appear intelligent without needing to be conscious. A salamander will sit patiently on a rock until an insect buzzes past, at which point it shoots out its tongue and captures its meal. This is an impressive ability but, just like a Facebook chatbot, the salamander does not have to be conscious or aware to do it. It is like driving a familiar route, such as your journey from home to work, which you can do automatically while thinking about unrelated things. There are people who suffer from the sleep disorder parasomnia who have been known to drive their cars while sleeping. Unconscious actions can be complicated and impressive while still not needing conscious awareness to carry them out.

For a long time, it was thought that only humans possessed consciousness. In the seventeenth century René Descartes argued that while animals could perceive the world and feel sensations such as pain, their actions were ultimately mechanical because they lacked the non-material essence of the divine human soul. The idea that animals did not possess a soul was a morally useful one, because it made it okay to kill and eat them.

To give Descartes his due, animals do seem to lack some conscious quality that humans possess. For example, if you watch a cowboy film where the cowboy rides up to a saloon and ties up his horse before entering the bar, you'll notice that the horse appears in no way concerned by this. There is nothing in its behaviour to suggest it is troubled by the awareness that it has been condemned to do nothing but stand on the spot for many boring hours. If you were to tie a human up outside a saloon and leave them there all afternoon, they would act very differently.

We often don't notice this missing quality in animals because we project human qualities onto them, especially our pets. The Canadian author Douglas Coupland calls this process 'bambi-fication', which he defines as the 'mental conversion of flesh-and-blood living creatures into cartoon characters possessing bourgeois Judeo-Christian attitudes and morals'. The notion that a particular animal is a 'good dog' or a 'bad kitty' comes very easy to us. It can be pleasing to interpret automatic animal behaviour in terms of human personality.

Yet the more animal behaviour is studied, the more we find examples of animals, particularly mammals and birds, exhibiting traits that do suggest some form of consciousness. Chimpanzees, elephants and magpies can recognise themselves in a mirror. This suggests that they possess self-consciousness and understand themselves as a particular individual. Some animals can communicate with their peers, appear capable of suffering, and can learn to use tools. Is this sufficient evidence to conclude that some of the more advanced animals are conscious after all?

The big problem here is that we can't define what consciousness is. As the academic textbook *Human Brain Function* admits, neuroscientists 'have no idea how consciousness emerges from the physical activity of the brain and we do not know whether consciousness can emerge from non-biological systems, such as computers [...] At this point the reader will expect to find a careful and precise definition of consciousness. You will be disappointed. Consciousness has not yet become a scientific term that can be defined in this way.' The *International Dictionary of Psychology* sounds a little tetchy when it describes consciousness as 'a fascinating but elusive phenomenon; it is impossible to specify what it is, what it does, or why it evolved. Nothing worth reading has been written about it.'

Can anything non-human be conscious? The lack of a definition of consciousness makes this tricky to answer, but it is still a question worth considering because of the implications for AGI. It can be enlightening, at this point, to consider the octopus.

From a human perspective, octopuses are about as weird as life gets. For a start, they can squeeze their bodies through a hole as small as their own eyeball. They have three hearts and blue-green blood. Their oesophagus, which connects the mouth to the stomach, runs through the centre of their brain. They also have twice as many brain cells in their arms as they do in their central brain itself, and they have the strange ability to tweak or edit the genetic instructions encoded in their DNA.

You have to go back a long way – about 600 million years – to find a common ancestor of both humans and octopuses. This was long before life had crawled out of the seas and before there was any such thing as invertebrates and vertebrates. Our common ancestor would have been a tiny wriggling worm-like thing, only millimetres in size. It would have been sufficiently evolved to contain some neurons, and possibly a pair of light-sensitive patches that were the beginnings of eyes, but it was otherwise very far removed from the mammals, birds, lizards and cephalopods that its offspring would become. Because humans and octopuses have been on a separate evolutionary path for so long, researchers studying octopus DNA say that it is the closest thing we can get to studying alien DNA.

If octopuses possess consciousness, then it has evolved entirely separately from our own. This would mean that consciousness is not some miraculous quality that only humans can possess. It would confirm that consciousness is indeed something that pops into existence when evolution has produced a sufficiently suitable brain. This, in turn, would make the idea that computers could become conscious more plausible.

Out of all the non-human animals on the planet, octopuses have the strongest case for possessing a form of consciousness that, while not exactly the same as human, is sufficiently similar for them to both be classed together. It is not, admittedly, easy to be sure. Octopuses don't play well with standard animal behaviour tests. For example, the Harvard scientist Peter Daws performed tests on octopuses in Naples in the 1950s, including seeing whether an octopus could be trained to pull a lever in order to release a morsel of food. Two of Daws's three octopuses duly pulled the lever and obtained the food. The third octopus, named Charles, also passed the test, but in something of a begrudging manner. He anchored his tentacles on the side of the tank so that he could apply great force to the lever. In this way he repeatedly bent the lever, and eventually succeeded in breaking it off.

Charles the octopus did not seem happy to take part in these experiments. As Daws wrote, 'Charles had a high tendency to direct jets of water out of the tank; specifically, they were in the direction of the experimenter. The animal spent much time with eyes above the surface of the water, directing a jet of water at any individual who approached the tank. This behaviour interfered materially with the smooth conduct of the experiments, and is, again, clearly incompatible with lever-pulling.'

Charles's ornery personality, although extreme, would probably not surprise people who have worked with octopuses in laboratory conditions. Octopuses come with many differing personalities, but they are antisocial creatures in the wild and they do not play nicely with others. They are also very good at recognising different humans, even when those humans wear scuba masks or identical uniforms, and they often take against some people and not others. The philosopher Stefan Linquist had trouble when he worked with octopuses because they would deliberately plug the outflow valve of their tank with their arms, raising the water level and ultimately flooding the lab. As he explained, 'When you work with fish, they have no idea they are in a tank, somewhere unnatural. With octopuses it is totally different. They know that they are inside this special place, and you are outside it. All their behaviours are affected by their awareness of captivity.'

When the BBC filmed a shark attacking an octopus off the coast of South Africa for an episode of the TV series *Blue Planet II*, they were amazed to see the octopus deliberately inserting its arms into the shark's gills. Unable to breathe, the shark had no choice but to break off the attack. The octopus then covered itself in shells from the sea floor and hid from the shark. None of this could plausibly be dismissed as intelligent-seeming learnt or automatic behaviour. It appeared that the octopus was aware of exactly what it was doing. As the camera operator Craig Foster later wrote, 'watching her trying to outwit a deadly catshark was terrifying for me [...] I totally fell in love with this octopus.'

Of course, octopuses have not demonstrated intelligence to

the extent that we have. They have not invented opera, or built cathedrals, or launched a space programme. But octopuses are relatively antisocial and the development of human culture was fuelled by co-operation. As the American biological anthropologist Agustín Fuentes writes, 'Countless individuals' ability to think creatively is what led us to succeed as a species [...] This cocktail of creativity and collaboration distinguishes our species – no other species has ever been able to do it so well.' Being boneless sea dwellers with a lifespan of only a couple of years, octopuses have little use for opera and cathedrals. But if they learnt to co-operate and communicate, are they sufficiently aware and conscious to be able to produce their own examples of intelligence and culture?

5.

The reason why awareness or consciousness, or something functionally similar, becomes an issue in the move from AI to AGI is because of the importance of original goal setting and defining purpose.

There doesn't seem to be any need for consciousness in learnt, mechanical activity that reacts to immediate stimulus, such as that demonstrated by the insect-catching salamander mentioned earlier. This can be sufficient to get a creature around its environment, find food, mate, and so on. If you shut down your higher brain functions by binge drinking in town, you usually find that you are able to get around, dance, interact socially, and obtain drinks and fried chicken. You will even find that you somehow make it home.

The Australian philosopher Peter Godfrey-Smith, whose bestselling book *Other Minds* discusses octopus intelligence, has written about the point where consciousness as we understand it evolves from the unconscious animal mind. As he sees it, 'The senses can do their basic work, and actions can be produced, with all this happening "in silence" as far as the organism's experience is concerned. Then, at some stage in evolution, extra capacities appear that do give rise to subjective experience: the sensory streams are brought together, an "internal model" of the world arises, and there's a recognition of time and self.'

Godfrey-Smith, inspired by the views of the French neuroscientist Stanislas Dehaene, goes on to discuss what we can and can't do without consciousness. He writes, 'We can't unconsciously perform a task that is novel, rather than routine, and requires a series of acts, step-by-step. [...] There's a particular *style* of processing – one that we use to deal especially with *time*, *sequences*, and *novelty* – that brings with it conscious awareness, while a lot of other quite complex activities do not.'

Here 'novelty' refers to something that has not been encountered before, and which could not be predicted based on past experience. Something mechanically reacting in the moment to learnt stimuli cannot respond to novelty. It requires an internal mental model of the world to make sense of the unexpected. This internal model seems to be an integral part of consciousness, and it makes it possible to think in terms of sequences of actions taking place over periods of time. The arrival of this 'internal model' of the world in the human mind appears to be what caused the cognitive revolution and marked us as different to the rest of the animal kingdom.

The need to plan, and to understand novelty, sequence and time, is the reason why DeepMind's AI failed at playing *Montezuma's Revenge*. It is also exactly what Eric's AlgoHiggs lacks. This code can create grammatical sentences, but it cannot string those sentences together in a way that says something interesting about the world, because it has no internal model of the world to describe. Without an awareness of what it is writing about, it is never going to be able to say anything genuinely interesting.

Consider, once again, our hypothetical AI which is being fed constant, real-time information about the city of Paris. When we imagine that scenario, we do so with the benefit of an internal mental model of the city. Much of our model may be vague, especially if we have never lived there, but we all know that Paris is a major European city with millions of inhabitants and that it includes the Eiffel Tower, the Louvre and the River Seine. When we imagine ways to put this Paris AI to good use, we need our mental model of Paris to come up with ideas. Perhaps the AI could take control of the traffic lights in such a way as to improve the flow of traffic, or to make travelling safer for pedestrians and cyclists. Perhaps the AI could monitor the position of police officers and ensure they are distributed in such a way that any disturbances could be reached quickly, regardless of where they took place. Or perhaps the AI could spot pickpockets and track their location until they are apprehended.

Imagining those suggested goals for the AI is made possible by using our internal mental model of the city, however rough and ready it may be. The AI lacks such a model, so it has no way of imagining useful goals. To the AI, all of Paris is nothing but a pattern of numbers. There would be no qualitative difference between the numbers generated by the traffic flowing smoothly and the numbers generated by unmoving traffic in a city-wide gridlock. Without a conscious human who is aware of the reality of Paris to set goals and tell it what to do, there is no way the AI could decide what would be a positive goal and what would be negative. AI relies on a human teacher to educate it about which patterns of numbers are desirable and which are not.

This idea that awareness needs an internal mental model is what makes octopus behaviour so important. Octopuses do seem to be able to plan actions in response to new and strange situations and environments. Their behaviour implies that they are capable of the style of thought that deals with time, sequence and novelty, and which requires their mind to hold an internal model of their world. Octopuses appear to have made the same leap in intelligence that would be needed to move from AI to AGI. Such a leap does seem indicative of consciousness as we understand it.

This is not to suggest that the experience of octopus consciousness would be in any way comparable to that of human consciousness. Lacking verbal language, octopuses presumably do not have the constant stream of chatter running through their heads that we do. Nor do we have a biology remotely like that of an octopus, which has a distinct brain in each arm. But there does seem to be a growing acceptance that a form of consciousness, as that mysterious term is generally understood, is the best explanation we have for the singular behaviour of octopuses. To a large part, the assumption of octopus consciousness has a lot to do with how being watched by an octopus feels considerably creepier than being noticed by a fish.

In 2012, neuroscientists at a Cambridge University conference signed the Cambridge Declaration on Consciousness, in which they asserted that certain animals were conscious. As they stated, 'The absence of a neocortex does not appear to preclude an organism from experiencing affective states. [...] Consequently, the weight of evidence indicates that humans are not unique in possessing the neurological substrates that generate consciousness. Non-human animals, including all mammals and birds, and many other creatures, including octopuses, also possess these neurological substrates.'

This has implications for animal welfare. On the one hand, causing harm to things that are living, but which are not conscious, is generally uncontroversial. I cut the lawn at the weekend, and in doing so sliced thousands of living blades of grass in two. I am untroubled by this and so far have not been criticised for it. If anything, I am criticised for not doing it enough. When Descartes declared that animals were not conscious, and the church declared that they did not have souls, then we felt no moral objections to keeping, eating and experimenting on them. Any dissenting voices could be dismissed as bambification.

Causing suffering to a conscious creature, on the other hand, is morally very different to harvesting mushrooms, stepping on an ant, or generally harming unconscious life. For this reason, seemingly unconscious invertebrates have historically been excluded from animal cruelty guidelines, although in a number of parts of the world those rules have now been extended to cover octopuses. Perhaps one day we will see church teaching extended to include the souls of octopuses. Octopuses can be little swines, and any church which attempts to look after their souls is going to have its work cut out.

If octopuses have been able to evolve a form of consciousness, completely independently from humans, then consciousness is not a miraculous, uniquely human attribute. This makes the notion of computers developing it more plausible. But it does not make it certain.

6.

It is possible that AI research will produce new insights into the nature of the mind. For example, there is a speculative branch of research that is interested in the idea of uploading human brains into computers. The idea is that if an individual's brain could be mapped exactly, down to the precise positions, connections and strength of connections for each neuron, then that brain could be modelled inside a computer. A digital copy of that individual, which thought in the same way and had the same personality and memories, would potentially be immortal. The assumption here is that if this ever became possible, then the model of the brain inside the computer would also be conscious. After all, if something was an exact copy, then it is logical that it would possess the exact same attributes as the original.

But there is a leap of faith at play here. Not understanding how consciousness arises, we can't say that a software model of neurons will generate consciousness just like a physical brain. On the contrary, there is a growing collection of research which suggests that it won't. For example, a 2013 paper by Stuart Hameroff and Robert Penrose claims that consciousness arises from quantum vibrations inside microtubules, which are a protein structure within neurons. In 2015, Matthew Fisher, a physicist at the University of California, published a paper in which he argued that a form of information processing occurs in the brain which is based on the quantum entanglement of phosphorus atoms. In 2017, Henry Markram at the Swiss Federal Institute of Technology in Lausanne argued that our brains create complex but ephemeral structures in at least seven dimensions when we remember, learn and think, and that consciousness may be 'a shadow projection from higher-dimensional representations'. It seems that we have not yet scratched the surface in terms of understanding what is going on. All these avenues of resear'ch are speculative, but they represent a growing recognition that the chemical reality of a mass of nerve cells is significantly more complicated than virtual neurons modelled inside a computer.

This is not to say that neurobiologists won't one day produce a complete picture of how a brain produces a mind, to the extent that we could model those processes accurately in a machine. It might require radical new computer architecture, such as the quantum computers which are currently in early development. Perhaps it will require a machine that utilises an experimental type of electrical component called a 'memristor'. A memristor remembers the amount of electrical charge that has previously flowed through it and adapts how it regulates the flow of electrical current in a circuit based on that memory. A computer built from memristors would, in theory, combine memory and processing in a way that was more brain-like than current neural networks.

Examples such as these are reasons not to dismiss the idea of conscious machines entirely, but we are clearly a long way from getting there. The idea that a quality akin to consciousness is just going to ping into existence inside a computer running a deep-learning algorithm in the near future is, at best, an expression of blind faith. There is no reason to think it is going to happen. It is not just that we haven't done the work needed to make it happen. It is that we have no idea how to even begin.

In films like the 2014 Johnny Depp thriller *Transcendence*, we are told that the useful AI tools we have now are suddenly going to evolve a soul, somehow become impossible to switch off, and then wipe out all of humanity. To the worshippers at the Church of the Singularity, the arrival of their AI deity is imminent. But if consciousness is necessary for generating an internal model of the world and choosing original goals beyond those triggered by immediate stimulus, then the evidence that AGI could emerge from our current AI techniques is basically zero.

For all that current AI is improving at an amazing rate, the probability that we're going to create an AGI is about the same as the probability that we're going to build a time machine. Like inventing time travel, it seems unlikely because nobody has any idea about how to go about it. That fact is usually drowned out by the excitement, horror and hype that surrounds the idea of an AGI. It is easy to be seduced by our habit of projecting personality onto machines.

7.

The usual method of predicting the future is to wait until a new invention or technology does something we've never been able to do before, and then claim that in the future this technology will be both common and better. Futurology like this can be called an arrow-flight projection, as it imagines a direct straight line from fledgling technology towards some super-powered, unimaginably impressive version of the same idea. This is a very fruitful method for anyone writing a science fiction story.

An arrow-flight projection is a bit like looking at a newborn male baby and declaring that, one day, the child will look as impressive as Dwayne 'The Rock' Johnson. Such predictions are usually over-optimistic. They can come true, of course. They came true for baby Dwayne 'The Rock' Johnson. But, more commonly, the growth and evolution of something are more idiosyncratic and unpredictable than the assumed direct straight line.

The best example of a successful arrow-flight projection must be Moore's law. Gordon E. Moore was a co-founder of the microchip manufacturer Intel. In 1965, he claimed that the number of components on an integrated circuit, and hence the speed and performance of computer chips, would double every year. This principle has proven to be accurate, even if the 'doubling every year' part has at times been tweaked to every two years or every 18 months in order to make the maths work. Belief in Moore's law and certainty about the rate of improvement has shaped the evolution of computers. Users, developers and manufacturers all expect and plan for rapidly increasing performance and shrinking costs.

Moore's law could not continue indefinitely, because the transistors cannot become smaller than the atoms they are made out of. It's now generally accepted that Moore's law is finally coming to an end, although access to massive data centres and the potential of future quantum computers means the end of the law won't mean the end of advances in computing. Moore's law complies with Stein's law, which, as we've already noted, tells us that 'If something cannot go on forever, it will stop.' It also complies with Davies's corollary, which states that 'Things that can't go on forever, go on much longer than you think.' When Gordon Moore made his original claim in 1965, he thought that it would hold true until the middle of the 1970s.

Because Moore's law has been so integral to the mindset of the computing industry, it has become a culture where arrow-flight projections are common. The belief in the AI singularity is a good example of this. As a general rule, however, things don't usually evolve so smoothly. When NASA landed Neil Armstrong and Buzz Aldrin on the moon in 1969, many arrow-flight projections were made claiming that mankind would walk on Mars by the 1980s or holiday in floating space hotels by the year 2000. These predictions didn't come true for the same reason that we don't have personal jetpacks, or travel in supersonic passenger aeroplanes like Concorde. The arrow-flight projections collided with real-world factors such as cost, safety, technical difficulties and demand. In the 50 years after the Apollo moon landings, no one was able to justify an attempt at manned Martian spaceflight. This was not because it was impossible, but because the reasons for going were not strong enough to overcome the cost and danger.

In *The Jetsons*, jetpacks were used by everybody, including small children and their pets. James Bond used one in 1965's *Thunderball* and another was featured in the opening ceremony of the 1984 Los Angeles Olympics. Even the visionary author Isaac Asimov predicted that jetpacks would be 'as common as a bicycle' by the year 2000. This was existing technology that we could see being demonstrated. It was easy to imagine it would get cheaper, better and more common. This didn't happen.

Our current lack of jetpacks has become a poster child for disillusionment about the future, as illustrated by the name of the Scottish indie band We Were Promised Jetpacks. Yet it is not true to say that we don't have jetpacks. They exist and can be bought online. The problem is that they cost around $200,000, hold enough fuel to fly only for a matter of minutes, and are insanely dangerous. The reasons why we don't all have jetpacks in the twenty-first century are that we don't need them and we're not suicidal. Once again, optimistic arrow-flight projections ran into the realities of demand, safety, technical limitations, cost and good sense. As Bill Suitor, one of the original jetpack pilots, explained, 'I hope they never become popular. Nobody would be safe. You'd have people falling out of the air like unwanted Santa Clauses. I've had several close shaves myself and almost sliced myself up like a big soft slice of silky cheese. Could you imagine every idiot who could afford one flying about?'

The impressive technical innovation needed to create an initial prototype is rarely the same problem that needs to be solved to turn that prototype into the imagined super-powered version of itself. Television was invented in the early twentieth century, for example, and this led to the *Dick Tracy* comic strips assuming we'd soon be able to speak to each other by video watches. This is something we can do now, but we had to invent digital video, mobile internet and the microchip to do so. Analogue television is a very different invention to these technologies, and its creation did not guarantee their development.

Emerging technology also has the ability to alter the territory ahead that it needs to cross, in ways that make simple linear predictions impossible. If we look back, we could claim that there was a direct, arrow-flight projection between the mechanical-powered looms of the Industrial Revolution and a modern automated car factory. Yet the inventors of the looms couldn't have made this prediction during the Industrial Revolution because they had no way of imagining cars. The steam-powered technology of the late eighteenth century changed the world in such a way that predicting what it would lead to became impossible.

When AI first appeared in the 1950s and 1960s, it came complete with exciting claims for its development and potential. In 1965, for example, the American computer scientist and polymath Herbert Simon said that 'machines will be capable, within twenty years, of doing any work a man can do'. Arrow-flight projections like these came from experts who were knowledgeable enough to appreciate the breakthroughs that had occurred in the field, but who were still unfamiliar with the difficulties that lay ahead. As a result, AI has a history of being oversold. The reality of what was achieved frequently differed greatly from what had been promised. This led to periods when the governmental or military agencies who funded research became disillusioned with AI and this, in turn, resulted in the money available for AI research drying up. In computer history, these are known as 'AI winters'.

The first AI winter came after the British mathematician James Lighthill produced a 1973 paper for the UK's Scientific Research Council. The Lighthill report reviewed the academic work into AI and was highly critical of the field's ability to achieve the 'grandiose objectives' that it set itself. According to Lighthill, promising prototypes repeatedly failed to mature into products capable of dealing with the messy complexity of the real world. The American defence agency DARPA concurred with Lighthill's conclusions, and an almost complete end to the AI funding in the UK, Europe and North America began in 1974. This first AI winter lasted for the rest of the 1970s.

In the 1980s, interest in the emerging field of expert systems saw research funding return. This benefited from the sudden increase in affordable computing power offered by the introduction of large numbers of processors working together in parallel. Once again, however, the real-world results did not match the optimism or ambition of the researchers, and a second AI winter began around 1990.

No one expects another AI winter to descend any time soon. Research is being funded by private companies now, and they are finding the fruits of that research to be profitable. Advances are coming thick and fast, and the rate of progress looks like it could continue for some time. Yet the historical trend of overselling AI's potential continues regardless. Journalists and CEOs alike have a vested interest in embellishing the potential of the technology. Some technology evangelists talk about future AI as though it were a God-like omniscient entity, so far in advance of us that we will never understand its purpose or its actions. But this idea, more than any other, makes the same mistakes that those who oversold AI in the early 1970s and late 1980s did. It assumes that we will find solutions to problems that we can't even define yet. This is the altar upon which arrow-flight projections usually crumble.

8.

Tales of conscious hyper-intelligent God-like machines may seem harmless, but dreams of what will never be distract us from the significance of what we are already creating. The impact that AI is having on our daily lives is already apparent and looks certain to increase. Perhaps the most interesting thing about this is that the tasks AI is good at tend to fall into two categories: things that we humans can't do ourselves, and things that we can do but find boring as all hell.

As an example of something that we can't do, consider the problem of identifying missing children who will now be years older than their last known photograph. When all you have is a photograph of someone aged perhaps seven or eight, it can be extremely difficult to correctly identify that person when they are in their mid-teens. This is a huge problem for those tasked with the fight against child trafficking. In response, researchers at Michigan State University have trained an AI on photographs of nearly 1,000 children, who were all photographed multiple times between the ages of two and 18. In this way, the AI is learning to recognise how individual children age. It has been able to correctly identify children 73 per cent of the time when they are three years older than their last known photograph. No one expects that such a system will ever be perfect, but it will continue to improve, and in doing so it will serve as a tool that we simply didn't have before. In circumstances like this, where a machine can do something valuable that we simply cannot, it is hard not to see AI as an asset.

Perhaps more commonly, AI is put to use doing jobs that bore us rigid. If you look at what AI is currently being used for, from assessing insurance forms to identifying images or maintaining armies of fake social-media accounts for nefarious purposes, you are unlikely to feel a pang of envy. The jobs that AI will be taking from us are not the kind of jobs that you'd dream of spending your life doing. There are exceptions, of course. Some people enjoy driving and will miss it if self-driving vehicles become the norm. But for the most part it is lucky that AI is not actually conscious. Such a self-aware machine would be bored beyond belief. If a machine knew that it had to spend 24 hours a day for the rest of its existence tweaking the order of news stories in a Facebook feed, it might consider this an act of cruelty to a sentient individual. If we did ever succeed in one day building a conscious super-intelligence, I would not be surprised if the machine's first act was to turn itself off.

AI is a reminder that the assumptions we unconsciously absorb from fiction are not the best guide to the world we're building. Yes, it will continue to evolve, and it will alter society in ways we can't see yet. But it will remain a tool, and those who use it will be responsible for it. There is no reason to believe that it will be able to set itself new and original goals. Here, then, is the first lesson we need to understand if we are to better understand the coming future: The creation of purpose shows no signs of being automated. That has been, and always will be, our job.

3.
PATTERNS

1.

On the first Thursday of every month, in the Victorian warehouses underneath Brighton Station, there is an experimental-music night called 'The Spirit of Gravity'. The entrance fee is £5, and this typically gets you three acts from the outer fringes of what we will loosely describe as music. There will be no songs, and no melodies. Instead you will be presented with ambient droning electronics, strange squawking noises, unsettling dissonance and occasional screaming.

The Spirit of Gravity is usually badly attended. It is rare that more than a dozen people turn up. The night originally began around the Millennium, so its continued existence shows a remarkable level of stubbornness. My artist friend Eric Drass loves it, for he has a particular fondness for heartfelt creative endeavours for which no apparent audience exists.

It is also incredibly well curated. Where the organisers find such a continual stream of uncompromising extreme musicians is a mystery. It is not unusual to find foreign artists on the bill who have travelled a great distance to appear. The night gives a glimpse into an otherwise invisible world, where experimental-music enthusiasts spend long hours of their free time in their bedrooms or garages honking, droning and spluttering for their own pleasure. It is noticeable that they usually arrive at the gig alone and don't bring friends with them, so I suspect they get little encouragement from their families. And yet they do it regardless. I can only assume that they do it because they love it.

Eric and I head to the Prince Albert pub to meet our friend Matt Pearson before the gig. Matt is a data scientist with a soft-spoken Wolverhampton accent, longish hair and a fondness for felt baseball caps. He works for a large data-analysis company. This company exists to harvest the vast quantities of information which people on social media freely broadcast to the world, in the belief that this will help large brands and companies understand the behaviour of potential customers. It's one way to make a living.

Eric is in two minds about Matt's job. On one hand he views it as extremely evil, and so is quietly impressed. On the other, he sees it as a huge waste of Matt's talents, which lie in the field of real-time generative art. Under the name Zenbullets, Matt codes visuals that are created on the fly in reaction to real-world events. He has written the leading textbook on generative art, and he used to code real-time projections for bands like Groove Armada. Because of these skills, Eric is not happy that Matt has taken a secure, well-paid job that allows him to pay his mortgage and raise his children.

I suspect Matt's choice of employment is not just practical. When he talks about the vast oceans of social-media data that he attempts to navigate, he talks about it like an explorer coming ashore on an unexplored land. This is new territory. It is hard not to be curious or intrigued.

As we settle into a quiet backroom of the Albert with our

drinks, I ask Matt why data is seen as being so valuable these days. 'It's partly for targeting advertising,' he tells me, 'but it's partly speculative as well.

'With companies like Facebook and Google, there are two levels to their share value,' he explains. 'One is based on their short-term outlook, and that is the advertising thing – how they can use what they know about people to target ads at those people more effectively. That is part of the reason for their share price, but that doesn't account for their insane share valuations. Those are actually more to do with the long-term speculative futures of these companies. They are building this corpus of data on the whole of the human race, and the market is speculating on what that is going to be worth in the far, far future.

'It's trying to factor in as-yet-unimagined revenue streams, but it's also kind of questionable. It's not certain that this data will ever be useable, or whether they could ever find meaning from that chaos in any way. It's a bit like – in our house we've got a drawer full of plastic forks out of ready meals. They were never thrown away, in the belief that we would get some use out of them in the future. For years we've almost filled a drawer with these little tiny plastic forks, as if plastic forks were going to be the new gold. We are yet to have that dinner party where everybody is served olives with a plastic fork to go with it. I may have to face facts, that those plastic forks have no value and keeping them is a waste of time.'

'Burn them all,' mutters Eric into his drink. It's not clear if he is following the conversation and giving his opinions on the forks, or just off in his own world and thinking out loud.

'Even assuming that a use for all that data could be found,' Matt continues, 'collecting the data is only part of it. Analysing it is the other part. The tools for analysing it have got to keep up. It might sound like that's going to be an easy progression, and that they are always going to keep in step with each other, but it's not really. The bigger the corpus of data, the harder it is to make any sense out of it. There's no limit to the amount of data being collected. Storage is the only issue and storage gets cheaper and cheaper. But the ability to analyse these large amounts of data does not necessarily evolve at the same rate. It gets increasingly complex the more data there is. There may be theoretical limits on what you can analyse.

'The problem with the data that we deal with at work is that it's a representation of people, but it's not those people. There's a massive gap between the data and what it is said to represent. We did a thought experiment at work: if you could take out one of our competitors, who would it be? The idea is to think about the strengths and the weaknesses of other companies. And I said the one company I'd like to completely take out of the whole sphere – and they're not a competitor directly – is Facebook.'

Eric looks up. The thought of Facebook being taken out has got his attention.

'And the reason I said Facebook,' Matt continues, 'is because the Facebook model of data has become the norm. It has corrupted the way people behave online. They feel that they have to play a game against the platform and they don't think that they can be themselves. They are altering what information they give out in order to game the system, to get likes and shares and to please Facebook's algorithms. If companies like Facebook were nicer, or if they did it with an open model where you owned your own data and it wasn't being exploited and sold to the highest bidder, then people would get over the privacy problem. The privacy problem is really a trust problem; it's that we don't trust these companies with our data.

'So, to go back to what we were talking about with the hypothetical value of far-far-future Google or Facebook based on them amassing the corpus of the twenty-first century, then you have to remember that on many levels that data is largely bollocks. You've got oodles and oodles of data of people presenting themselves in their best possible form, having a great time, posting their holiday snaps and things like that, having fun with the kids. But that's not the reality of the twenty-first century. There's no data on domestic disputes or miserable childhoods or chocolate bars or what life is really like.

'It's like Victorian photographs. Because photography was such a rare thing Victorians would go and pose for their photograph in their best clothes with their children standing still and nobody smiling – it's not a typical scene of what the average Victorian life was like. There's very little photography of the working classes in Victorian eras, and whenever some comes up it is always fascinating because that is the data we don't have. That's the gold. Or it's like Neolithic people, where all that has survived of them are bones and rocks, so we think that their entire lives revolved around bones and rocks even though of course it didn't. Facebook data is just that – it's Facebook data, nothing more, and it's absolutely not an accurate model of how people act in the world. It's certainly valuable in certain circumstances, but that's not to say that it has great future value.'

I think back to an incident reported by the American web consultant Eric Meyer. On Christmas Eve 2014 he logged on to Facebook and was presented with a preview of a feature called 'Your Year in Review'. The image showed a photo of his six-year-old daughter staring out at the viewer in a confident, knowing way. His daughter had died a few months earlier from an aggressive brain cancer. Facebook had surrounded her photo with happy dancing figures having a party, complete with balloons and streamers. Above this were the words 'Eric, here's what your year looked like!' In 2018, a woman named Anna England Kerr was bombarded by parenting ads after having a stillborn baby, even after changing

Facebook's settings to not show her any parenting ads. Facebook's timelines are so far removed from the realities of life that no one coding them realised that things like this would happen.

2.

It is time to head over to The Spirit of Gravity. The venue is a dimly lit brick Victorian warehouse, with exposed girders and rusting pipes doubling for decor and cobbled floors worn smooth over the years. We arrive in time to see a duo called Onin, who are described on the flyer as 'Guitar, sax, dynamic feedback improv'. This troubles Eric, who is fundamentally opposed to saxophones.

The band do not take to the stage, as you might expect. Instead, they hunker down in the remains of a room once built in the corner of the warehouse. This has now crumbled away into little more than a freestanding doorway with two remaining walls that are no more than a couple of feet high. The audience, such as it is, stands around in the gloom or sits on the stage facing this ghost room in the corner. The performance begins.

I had no idea how many non-musical noises two musical instruments could make. An acoustic guitar, it turns out, can sound like mice holding a political rally. A saxophone can sound like a car meeting its end in a wrecker's yard, much to Eric's relief. The duo use a wide range of electronic trickery, violin bows and a hand-held fan to create these extraordinary noises. At one point, an actual note was played on the guitar, although I think this was an accident. The almost entirely male audience watch in respectful silence, which makes things worse if you are overcome by a fit of the giggles.

After an unknown period of time, the performance winds down. When the audience realise this – which is not easy, given the extent to which long periods of silence were part of the performance – they break out into enthusiastic applause. We wander back into the bar.

'I really enjoyed that,' says Matt.

Would an AI recommendation engine have suggested that the three of us go to that concert, if it had been given data detailing the music we listen to?

Eric, Matt and I have differing musical tastes. Eric favours bands made up of angry, shouting young people performing in near-empty venues, Matt prefers abstract electronica, ideally performed by girls with fringes, and I freely confess that the gig I've enjoyed the most in recent years was Iron Maiden at the Liverpool Echo Arena. If you had fed this information into an AI recommendation algorithm it would have tried to split the difference and suggested that we go to see The Pretenders, who were playing that night at the Brighton Dome. This feels like it would be a safe suggestion. Nobody actively dislikes The Pretenders.

But we had no intention of going to see The Pretenders. Our gig-going has a complex social logic. Its primary role is to allow each of us to attend the type of events that would sorely test the patience of our wives and partners if we were selfish enough to try and drag them along. It is not the case that we always loyally support each other's gig choices, as a complex set of variables are also in play that decides if a gig suggestion is accepted. These favour gigs played in rooms above pubs, bands that we have never heard of but who have a good name, and events that cost less than a tenner to get in. If one of us suggested going to see The Pretenders, the notion would have been rejected because they tick none of those boxes.

What would an AI make of data about the gigs we have attended? A computer could easily store a list of the dates, venues and bands involved in our gig-going. Our complicated and often contradictory reactions to the music played could be reduced down to a manageable data unit, such as a score out of ten. From that, you might imagine that AI could predict which future gigs we would enjoy. Yet the social, male-bonding aspect of our attending these gigs, and the fact that we can enjoy not liking the music we go to see, are the kind of details that computer data struggles to record. Data is a store of what can be easily captured about our lives, but this is only a fraction of the true picture.

AI-powered music-recommendation software, such as those used by services like Spotify, have a record of all the songs I have played in the past, and it knows how often I have played each song. But it doesn't know why I have played those songs, or the associations I have with particular pieces of music. I have fond memories of the Queen song 'Don't Stop Me Now', for example, which date back to when I was still in school. On occasions when the school music room was empty, a few of us would sneak in so that our friend Alan could play the piano, and he would invariably play 'Don't Stop Me Now'. Because of this, no matter how old I get, I will never be able to think badly of that brilliantly ludicrous song. None of this context is captured in the data used by AI. Every now and again I play that song on Spotify, just to check that it is still the glorious expression of teenage joy that I remember. But the AI recommendation algorithm does not understand why I play it. Instead, it works on the mistaken assumption that I like listening to Queen, or to songs that sound similar to 'Don't Stop Me Now'.

Based on my playing of 'Don't Stop Me Now', Spotify created a playlist of songs that it thought I would like. This included a number of songs by Rush and Foreigner, which in truth I could live without. It also suggested the song 'Mr Apples' by Madness, possibly because it is taken from their similarly named 2016 album *Can't Touch Us Now*. In a comparable way, the recommendation algorithms used by Netflix know what you have previously watched, but they don't know if you loved that programme or if you got bored and distracted and left the room. It doesn't know that it continued to auto-play further episodes for the benefit of nobody but the cat.

'The thing with AI recommendation engines,' says Matt, 'is that they are always going to draw you towards the norm. If you are different in your own special way – as all us special snowflakes are, we all have our own little foibles – then Amazon or Spotify or whichever recommendation algorithm is trying to serve you new stuff is not going to recognise those little foibles. The algorithm is always going to be addressing the commonality you have with the rest of the human race, not the thing that makes you ... you. It's like in *Citizen Kane*, if we whisper "Rosebud" with our dying breath, then the AI is not going to understand why. It's going to suggest getting some roses for the funeral, on the grounds that everybody likes roses.'

This tendency of recommendation algorithms to draw you towards the norm became apparent as music streaming established itself. There were 45 billion songs streamed in the UK in 2016, but fewer new songs made it into the Top 10 than in previous years. Only 11 songs made it to number one, compared to 24 in 2015 and 38 in 2014. Big hits are getting bigger, and everything else is falling away. This is great news for Drake and Ed Sheeran but worrying for up-and-coming artists. The music industry is coming to resemble Hollywood, which focuses on blockbuster hits with wide demographic appeal and built-in brand awareness. Anything original or challenging is not going to do well in this world.

Recommendation algorithms drag us all to the centre, to the bland and the safe, because it's just a numbers game to them. A living vibrant culture, however, is anything but a numbers game, and to reduce it to one, because of how limited the available data is, is to miss the point quite badly.

This raises the question of whether what is gained from the system is valuable enough to outweigh its flaws. Is the fact that we as a society have shared communal touchstones such as Ed Sheeran and Marvel superhero movies more important than the unthreatening quality of those touchstones?

It is time to head back to The Spirit of Gravity to see the next act. They are called Well Hung Game, and are described as 'Baritone sax and dirty electronic processing'. Eric is far from happy about the mention of baritone sax, but we head in regardless.

Following the recommendations of algorithms is not compulsory. They are trying to help and they are doing their best to introduce us to new things that we won't hate, but they are a long way from grasping the complex social considerations that drive us. We are free to ignore them, and we should make a point of doing so every now and again. After all, by the logic of the market, by rational consideration and by common sense, The Spirit of Gravity should not exist. But exist it stubbornly does.

3.

In an effort to not be entirely sedentary, I started wearing a fitness tracker on my wrist. This little device measures my steps, sleep and heart rate, and it reports them faithfully to me via an app on my phone. This data is not just a passive record of my life. It changes what I do and how I act. Many of these changes may be small, but they add up. I am now more likely to park at the far end of a car park, for example, because I know this will force me to walk further. The knowledge of how little I have moved all morning pushes me into taking a lunchtime stroll that I would otherwise not think of taking.

There are times, however, when it fails to record my steps. If

I am pushing a shopping cart around a supermarket my wrist remains still on the trolley's handle, and none of my steps are recorded. Sometimes I take the wristband off to charge, forget to put it back on, and head out for a walk without it. The app then reports that I have taken fewer steps than I really have. I react by forcing myself out for another walk so that the app can tell me I have walked the correct daily amount. I do this knowing full well that in the real world I have already walked the required distance to meet my target. I know that the data is wrong and that my goal has been met, yet I defer to the false data over the truth of the situation. My memory of the truth will soon fade, but the data will last. This is, I think, quite common behaviour among those who wear fitness trackers.

Walking out of The Spirit of Gravity and making an ill-advised late-night visit to the whiskey bar at the bottom of the hill, I continue to discuss the behaviour of algorithms with Matt. 'I'm really wary of those situations where you feel that you are compromising your human decisions for something that fits into an algorithmic view of the world,' he tells me. 'I get told off by my meeting-room booking system at work. It nags me. I really resent myself for actually being affected by that. It's just an automated message with no emotion behind it, but I feel like I'm being told off. It says, "You've booked this room, you're fifteen minutes late, I'm going to cancel it. Please stop wasting people's time", or something like that. And I know that it's just an algorithm doing this, there is nothing personal about it. But I find that I'm changing my behaviour to keep that algorithm happy. In theory the algorithm is designed to help humans, but what's actually happening is that humans are bending over backwards to help the algorithm.'

The Israeli historian Yuval Noah Harari defines 'dataism' as an emerging religion that venerates neither gods nor man, but which instead worships data. As he defines the term, 'Dataism declares that the universe consists of data flows, and the value of any phenomenon or entity is determined by its contribution to data processing.' In this view of the world, only data exists. The records of David Bowie, the recipe for cheesecake, the rules of football or the Microsoft corporation are all understood as data being processed. Living organisms are considered to be algorithms. Bacteria, octopuses and human beings are all methods for processing data. So are political institutions, the stock market and the global climate.

Harari is not arguing that dataism is a good thing, or that we should adopt it. He is aware that dataism could mean 'adopting an increasingly skewed view of life' which disregards aspects of existence not relevant to data processing. He writes that 'Dataism is neither liberal nor humanist. It should be emphasised, however, that dataism isn't anti-humanist. It has nothing against human experiences. It just doesn't think that they are intrinsically valuable.' His point is not that dataism is a utopian or dystopian future. His point is that dataism has become our default method for understanding the world, without us really noticing.

Looking around, it seems that he has a point. My deferring to the fitness tracker's data and Matt's attempts to please the meeting-room booking algorithms all point to this. I am old enough to remember how film fans discussed movies back in the twentieth century. The notion that movie lovers should give any consideration to, or indeed even know about, the amount of box-office money a film made was then pretty much unheard of. Even if people knew the figures, how were they relevant to anything important or valuable about what matters, namely the subjective experience of watching the film? Now film enthusiasts not only pay attention to data like a film's domestic and international gross, but they consider these unarguable markers of a film's worth. Because this one aspect of cinema lends itself to objective data processing, it has been radically elevated in importance.

We all know that these are only numbers. They are a sampling of what can be sampled and say nothing about what can't be reduced to data. As the British Zen philosopher Alan Watts used to stress, the map is not the territory and the menu is not the meal. On an aggregate review data website such as Rotten Tomatoes, many differing and varying film reviews are aggregated to produce a final number, a percentage figure between 1 and 100, that stands as an objective scoring of a film's quality. Under this system, a film that was passionately loved by some and hated by others will receive a similar, mid-table aggregate score to a bland, forgettable, unambitious movie. Movie lovers and studio heads alike all know this, yet they obsess over Rotten Tomatoes scores regardless.

My partner, Joanne, also has a fitness tracker, but hers was made by a different manufacturer. Whenever we went for a walk together she would report having taken more steps than me, which I assumed was the result of me having longer legs. The strap on my tracker broke, however, forcing me to replace it, and I chose the same model that Joanne had. With this new tracker on my wrist, I suddenly found I was recording noticeably more steps than I previously had over familiar walks. I much prefer my new tracker for this reason, but I can no longer see its claims about the number of steps I have taken as some unarguable truth. I presume that one manufacturer is more accurate than the other, but how can I know which? Perhaps both are equally inaccurate, just in different directions. The tracker also gives a different number for my heart rate than the heart rate monitors built into equipment at the gym. You might have thought that measuring heart rate, which is basically just counting, would be the sort of thing that machines could agree upon, but it seems not.

Back in the 1980s, the Top 40 music chart was a big deal. Millions would tune in to BBC Radio 1 every Sunday to hear the countdown and find out whether their favourite singles had gone up or down. Then, in 1984, independent radio launched a rival programme called *The Network Chart Show*, which had its own, separately generated music chart. In this new countdown, the positions of songs did not always match those of the official BBC chart. As a teenage music fan at the time, I found this oddly disturbing. I had no sense of being pleased that I now had more data, just the troubling realisation that the positions in the official chart were actually far more arbitrary than I had realised. As the old saying goes, if you want to be sure, buy an encyclopaedia. If you never want to be sure again, buy two encyclopaedias.

Susan Bidel, a senior analyst at Forrester Research in New York, talked to the BBC in April 2018 about the huge amount of personal data being traded by data brokers. She claimed that a common belief in the industry was that only '50 per cent of this data is accurate'. Fake names, dead email addresses and mistyped phone numbers saturate the information that data brokers sell. A truism used in the data industry is that faith in data grows in relation to your distance from the collection of it. A statistic always sounds more convincing the less you know about where it came from.

'Another problem is that the data, and particularly social data from social networks, comes from quite a narrow demographic,' says Matt. 'It's not representative because it's not everyone. It's people of a certain wealth and it tends to be skewed towards the left, the young and the tech savvy. So if you look at a political party like UKIP that appeals to the opposite demographic – well, they're not invisible on social media, but they are certainly under-represented.

'I've done a lot of work these past years failing to use social data to predict elections. At best, I've managed to fudge the figures to get kind of close, but it always involves a lot of tricks and making excuses. And that's from using the entirety of the British social media as a data set – it's masses of data but the quality of the data isn't good enough. And there's also the problem of bots. When you're analysing social data, it's not always easy to know how much of that data is real and how much of it is fake networks of propaganda bots.'

'It's very hard to create neural networks that aren't biased,' added Eric, 'because it's almost impossible to generate data sets without bias to train them on. AI can only represent the data that you give it – all you're getting back is a statistical representation of the world as you've shown it to them.

'Remember when Google was using image-recognition AI to automatically tag photographs and it began tagging black people as gorillas? The problem was not so much the software but the photographic data it had access to, because photographic technology favours white skin tones. Kodachrome, the colour film that came out in the 1980s, was effectively the first film that showed black people with any level of accuracy. If you look at photos taken on colour film before then you'll see that black faces were pretty much just black, because of the chemistry of the film emulsion. Colour film was developed by white people and sold mainly to affluent white people who could afford cameras.

'The horrible thing is that this wasn't the reason why they changed they film chemistry. It was because furniture manufacturers in the States complained that they couldn't tell the difference between the different kinds of woods in their catalogues. They wanted a film that could show the difference between teak and oak. Kodak developed a film that would give them this, and that went on to have a massive effect on the representation of black people in society. It helped correct a major failure of our representation of society that had never occurred to those in charge. Or to give another example of data bias, there was a chatbot called Tay that Microsoft put online, and after 24 hours of learning from Twitter users it was tweeting that feminists should burn in hell and that Hitler was right, at which point they pulled the plug. That was an example of bias being deliberately introduced into a system, just because one group of people thought it was funny. So when you think about how biased society is in all sorts of ways, you see how seemingly impossible it is to get unbiased data with which to train AIs.'

I start to wonder if we value data not because it is true, but because it is decisive. Life is a messy subjective experience, but a single piece of data can be grasped objectively and agreed on by many different people in an instant. It forms common ground from which they can proceed to build. It is easier to accept what the data says, even when you know how arbitrary it might be, than to find a truer picture of the world.

Data only records particular aspects about the real world. These are frequently not the important aspects. We treat it as objective fact, even though it is full of bias and inaccuracies. It only captures what is easy to record, not what is important. None of this is news to data analysts. So why is data so valuable in the modern world?

The reason is because something unexpected happens when the size of the corpus of data becomes sufficiently large. The properties of big data are not the same as the properties of individual pieces of information. When you take a large enough perspective, data reveals patterns that were not apparent at a smaller scale. The process is similar to how the circumambient mythos appears out of an accumulation of all the individual cultural narratives, provided you can get sufficient distance to see it.

As the common truism in the data industry explains, more data is better than good data. The theory is that, yes, some data points will be wrong, but given a big enough data set these errors will cancel themselves out, allowing us to confidently draw conclusions from the database. Inaccurate data, if collected over time in massive quantities, can still reveal patterns and changes in aspects of the world.

It is these patterns, rather than the data itself, that is the gold. AI is essentially a pattern-recognition machine. It is able to see patterns in massive data sets that humans could never find. It helps us transcend the human-scale perspective and makes us think in terms of much larger, complex systems. Big corpuses of data tend to be trusted even when we know that individual pieces of information are flawed, especially if we can recognise and compensate for the social biases present in that data. It is still sensible to treat single objective statistical statements derived from data analysis with a pinch of salt, but the overall patterns revealed, and the changes to those patterns across time, can be extremely illuminating.

These strange qualities of big data can be difficult to accept. Big data, like statistics, is not something that we evolved to use or understand. Our inbuilt biases intuitively prefer a human-scale perspective and reject the idea that, when it comes to data, quantity is more important than quality. Yet, this does appear to be the case. That much of the data is garbage does not mean we cannot learn anything from a big enough pile of it.

4.

Big corpuses of data are how AI understands the world. While this is obviously very different to the 'internal models' used by conscious human minds, there are some interesting similarities.

A useful metaphor for understanding our internal models of the world comes from the American Harvard psychologist Timothy

Leary. This is not, admittedly, what Leary is best known for. In the 1960s, this once-respectable academic psychologist became the global figurehead of the psychedelic counterculture. His blatant promotion of the use of drugs such as LSD and psilocybin, through the slogan 'Tune in, turn on and drop out', led to him becoming, in the words of the Nixon government, the 'most dangerous man in America'. Leary was arrested and imprisoned, and broke out of his Californian jail with the aid of the radical terrorist group the Weathermen. His life repeatedly swung from high adventure to ludicrous farce, in the full glare of the global media, while surrounded by luminaries such as John Lennon and William Burroughs, and all the while keeping up a particularly hardcore regime of personal psychedelic drug use. It is perhaps understandable that, given the rest of his life, his ideas about mental models have been a little overshadowed.

Leary argued that each of us lived in our own 'reality tunnel'. As he defined it, your reality tunnel is not reality. It is what you think reality is. For most of history, we lived inside our own reality tunnels and mistook them for the world beyond.

Over the course of a lifetime the mind builds an internal model of the world based on our senses, experiences and memories. This model is extremely useful because it allows us to contemplate things that have happened and to make predictions about what may happen next. It allows us to understand time, sequence and novelty, which are the same processes that the neuroscientist Stanislas Dehaene identified as requiring consciousness.

But, as Leary would stress, our reality tunnels are only models. Models are by definition a simpler, less detailed representation of something considerably more complicated. They are not an exact one-to-one match, and a reality tunnel will differ from true reality in many areas. The problem is that we don't know exactly where, because our consciousness only knows our reality tunnel and not reality itself.

The belief that reality is exactly as you perceive it is known as 'naive realism'. This belief has been criticised since the days of Plato and the Buddha, but it is still an easy habit to fall into because our reality tunnels are so convincing. As a result, we tend to assume that people who see the world differently to ourselves, because they have a different reality tunnel, are either crazy, wrong or heretics.

The concept of reality tunnels reminds us that we are wrong about things far more often than we usually recognise. We only have one perspective on events, and that may be misleading. You will never meet anyone who agrees with you on every subject on every level. The way we fool ourselves into avoiding the implications of this is to subconsciously assume that everyone else is wrong and doesn't 'get it' like we do. This is the blind spot in our thinking that the reality tunnel model aims to bring into focus. It reminds us that it is not just everyone else who gets things wrong. It is everyone, yourself included.

Reality tunnels behave as if they are attempting to defend themselves. They don't like to be contradicted. Leary argued that when they encounter a piece of data that reinforces how they understand the world, they will typically accept it without question. This process is now more commonly referred to as 'confirmation bias'. There is some evidence that our susceptibility to confirmation bias is linked to the production of dopamine in the brain's frontal cortex, implying that our brain receives a positive hit of that reward chemical whenever we encounter anything that fits in with our prejudices. The algorithms that control what we view on social media are designed with this in mind, as we shall see later.

In contrast, whenever we encounter evidence that contradicts how our reality tunnels understand the world, our natural reaction is to ignore or deny it. This process is known as 'observer bias' or the 'observer expectancy effect', and it can lead to false beliefs becoming stronger, even in people provided with evidence that they are wrong. It's interesting to note the similarity between how these biases work and the method of training neural networks. These are also rewarded for producing a result perceived as being correct and dismiss results perceived as wrong, regardless of how biased or flawed the data is.

A great deal of work has been done to try and explain why we have these biases. Many researchers point to the importance of sociability in our early evolution because, from a survival point of view, it was often more important to be in sync with the tribe than it was to be correct.

Reality tunnels are a nice metaphor for how our conscious inner model of the world works, but they are also a metaphor that helps explain how AI 'understands' the world. In this model, the large corpuses of data that AI is fed with are that AI's equivalent of a reality tunnel. Data is all that the AI knows about the world, and it can't immediately tell the difference between accurate and inaccurate date. This data, of course, is less accurate than we like to think, and it is also the product of biases. AI can run error-detection algorithms on its data, just as we can test our beliefs and assumptions, but it can never know how much bad data those algorithms miss. AI will at times be wrong about how things are, in ways that it is unable to detect, just like a human can.

5.

A good example of how AI is not always right is state-of-the-art facial-recognition software, as used by the FBI and the police. This is said to be 95 per cent accurate. It is impressive, time-saving technology, but 5 per cent of false identifications is still a significant number of people. There are demographic biases that make the software less likely to identify women, black and younger people. It is also based on the assessment of mugshots, which are usually clear and well lit. Images taken from CCTV are considerably more difficult for the system to identify.

South Wales Police experimented with facial-recognition software in real-world scenarios, and from May 2017 to March 2018 the technology flagged up 2,685 individuals as people of interest. 2,451 of these proved to be false alarms. In London, the facial-recognition system used by the Metropolitan Police only correctly identified two people, according to information released under Freedom of Information laws in May 2018. Neither of these people was a criminal, and no arrests were made. The Met's technology had an alarming false positive rate of 98 per cent, which raises a great deal of doubt as to how effective the facial-recognition system used in China actually is. In November 2018, a photograph of the businesswoman Dong Mingzhu was shown on a giant Billboard-sized screen in Ningbo to publicly shame her for jaywalking, but she was entirely innocent of the offence. The AI facial-recognition system had failed to realise that what it thought was her in the middle of the road was just a picture of her on the side of a bus.

The software will get better as it learns from more data, but it will make matches that are false positives and it will fail to recognise faces that are in its database. False positives can be reduced by making the system less sensitive, but that can make it easier to miss what it is looking for and this can often be more serious than an overcautious mistake. When an Uber self-driving car killed the pedestrian Elaine Herzberg in Arizona in March 2018, the car's software had spotted her, but it did not brake because it decided she was a false positive.

AI can also be gamed or hacked. An object or image can be

designed in such a way that it becomes a kind of optical illusion to visual-recognition systems. AI researchers call these 'adversarial images', and they are usually created to prevent people being recognised by facial-recognition software. In 2017, a team of students at MIT researching adversarial images created a 3D-printed toy turtle. If you or I were to pick that turtle up and look at it from all angles, we would say that it was nothing more than a turtle. The only thing that might seem a little unusual is that some of the patterns on its shell look a little smeared, but we would only notice that if we were looking out for it. When Google's AI looks at the turtle, however, from almost every angle, it identifies it as a rifle.

The students needed to understand how Google's visual AI system worked in order to game it in this way, but if that knowledge is publicly available it is possible to work out how to make any object register as any other object. Consider what could happen if someone handed out free T-shirts at a public place where an AI visual-recognition system is monitoring the CCTV, such as an airport or a theme park. To our eyes, those T-shirts might look as if they have nothing more controversial than a logo or a picture of a popular character on, and we would think nothing of wearing them. To the AI, they could appear as bombs or assault rifles. You can imagine the chaos that would ensue.

A particularly inspired way to fool AI facial recognition was demonstrated by researchers at Fudan University in Shanghai. They developed a baseball cap with infrared lights under the brim that shone down on the face of anyone wearing the hat. Although the infrared light projected onto the wearer's face can't be seen by the human eye, it successfully fooled facial-recognition software. The researchers designed the hat in such a way that anyone wearing it was identified by AI as being the vegan musician Moby.

The errors of AI aren't always a serious problem. Even systems designed to do life-dependent work, such as self-driving cars or medical diagnosis, don't have to get it right all the time. They just have to get it right significantly more often than humans do. If a self-driving car runs over a pedestrian, then that is a tragedy. You can understand the anger of the family of the victim towards driver-less cars. But if the introduction of such cars also cuts the amount of traffic accidents and related deaths by a significant amount, then it is hard to argue that the technology shouldn't be used.

Some systems don't even need to be right very often. The AI that constantly makes recommendations about what it thinks I will like on platforms such as Spotify, Netflix or Amazon only rarely recommends anything that I have not heard of, but which turns out to be exactly what I was looking for. The aim of those recommendation engines is not to find things I will like, but to keep me engaged with the platform itself. It wants me to spend time exploring Netflix, Spotify or Amazon, and keep them as part of my daily life. This is a goal those systems fulfil admirably.

All this leaves us in quite an absurd situation. AI is essentially a machine that finds patterns, and it is vastly better at finding patterns than we are. But those patterns aren't in the real world. They are patterns in the AI's own reality tunnel, which is the data it has been fed. The AI never really knows where its reality tunnel differs from reality itself, even when sophisticated mathematical algorithms reduce the number of false positives and false nega-tives. It doesn't know if each individual bit of data is accurate, mistaken, or a sophisticated trick being played by academic fans of Moby. The hope is that all the mistakes will average out, and that the pattern will still be a good match for reality, even though we know that biases in the data make this far from certain. AI then offers up its patterns to us humans. We want those patterns to be an objective truth, and we are psychologically wired to regard them as such.

And yet, even with the flaws in the scenario apparent, the

patterns found by AI are still incredibly useful to us. True, they are patterns found in an abstraction of the real world and not from the real world itself, but we are still better off with them than without.

The failures of AI become problematic if there is no right of appeal. Imagine that an AI decided to turn down your application for a mortgage, and you were part of the small percentage of cases in which you are a valid client and the machine had made a mistake. In these circumstances, you would need the option of asking for a human to look again at your application.

Modern machine-learning systems are effectively 'black boxes', in that we can see what information goes in them and we can see what result comes out the other end, but we have no idea how that decision was reached. The many complex layers of ever-shifting neural nodes they contain are essentially opaque. We judge their effectiveness by analysing the results they produce, and should this exceed an acceptable limit we consider them safe for commercial use. But they can't be left running unattended. They still need to be monitored. They can be trusted with a lot of repetitive, time-consuming work, but they can't be given full responsibility for that work. AI could be of great help monitoring air traffic, for example, but the moment something unexpected happens, such as a flight disappearing from radar, then humans need to take the reins.

AI needs to be thought of as something like a tool or an assistant rather than a competitor that will 'take over', as the sci-fi dystopias portray it. As an example, consider the work that Google have been doing to assist research into the development of nuclear fusion. If mastered, nuclear fusion would provide us with abundant clean, carbon-free energy, with none of the problems of radioactive waste which accompany our current nuclear-fission technology. The problem is that nuclear fusion is incredibly difficult. It requires the creation of ultra-hot balls of gas called 'plasmas', which have proved to be notoriously unstable and short-lived. Billions have been spent developing the technology since it was first theorised nearly a century ago, but as yet it has stubbornly remained out of reach. Google's algorithms are designed to find new and better solutions to the problem, but they require a skilled physicist to guide them. 'We boiled the problem down to "let's find plasma behaviours that an expert human plasma physicist thinks are interesting, and let's not break the machine when we're doing it",' explained Ted Baltz from the Google Advanced Science Team. 'This was a classic case of humans and computers doing a better job together than either could have separately.'

AI will continue to improve. It will, as Turing foresaw, appear to be intelligent. It is going to have a profound impact on the world of work. Tasks that involve making complex decisions in the moment based on mountains of data, such as identifying tumours in X-rays or driving a vehicle, appear ripe for automation. We will have conversations with machines and find them valuable, and AI will become an increasingly prominent part of our daily lives.

The technology will not just be used for good. It will be used to cut corners in the pursuit of easy money. Some will avoid providing the right of appeal or human oversight that needs to always be present. AI will be put to tasks that are deliberately criminal, or just plain negligent. Someone, somewhere, is surely experimenting with adding AI to computer viruses. AI chatbots will prove to be more effective than spam emails at tricking people into giving out financial or personal data. The development, production and sale of weapons which decide for themselves if and when they should fire needs to be banned by international bodies, in much the same way that chemical weapons are. Organisations such as the Campaign to Stop Killer Robots, which lobbies on this issue, are highly necessary. Because AI is limited by the reality tunnel of its own data set, it can never be ethical to allow it to make the final call on life-or-death situations. Well-intentioned AI use will also, in some cases, have unexpected negative consequences that need to be recognised rather than glossed over. AI is a powerful tool that will be used for both good and bad, and legislation needs to keep up with its development.

But the idea that AI will become conscious or aware in a similar way to humanity is not supported by the evidence, or by our current technology. Nor should we assume that we can ever remove ourselves from a supervisory position and allow AI to 'take over', trusting it to make our decisions for us. AI is not something like Skynet from *The Terminator*, which plans to wipe us out, or HAL 9000 from *2001: A Space Odyssey*, which has goals above those known by the human astronauts. In practice, AI is more like R2-D2 from *Star Wars*, a useful assistant who is always there for you and will do as it is told.

For all the society-changing potential that AI offers, it is just as trapped by the limitations of its reality tunnel as the rest of us are. The quantity and quality of data will improve, but there will always be a gulf between data and the real-world event which that data is intended to represent. We humans have this flaw also, and we have learnt to muddle along with it as best we can. As history shows, it is a big flaw, but it is not fatal. We need to acknowledge it, however, because we have passed down our Achilles heel to our machine descendants.

6.

It may just be me, but I find that reading the latest news sections of science and technology magazines becomes addictive after a while. Every week new breakthroughs are announced, and knowledge that was previously occult and hidden enters the public sphere. Popular science magazines such as *New Scientist* or *Scientific American* reliably reveal promising breakthroughs in vital areas such as dementia research or nuclear fusion. These weekly and monthly magazines are a constant reminder of how resourceful, curious and inspired we humans can be. It doesn't take long, though, before you start to notice that those promised inventions rarely materialise as promised. Nuclear fusion still doesn't seem to be any closer than before. We remain incapable of curing dementia.

What usually happens is that new discoveries lead us to problems that we hadn't been aware of previously. The deeper and richer our knowledge, the more we understand how complicated things are. Sometimes, research flounders because of a shortage of money or the lack of an economic argument for further work. In other cases, rival scientists fail to replicate the original research. There is an issue known as the 'decline effect', which claims that the strength of scientific effects declines in a linear fashion after the excitement of an initial discovery, particularly in fields such as medicine, ecology and psychology. An example of this would be second-generation anti-psychotic drugs. Initial research showed these to be considerably more effective than first-generation drugs, but in time they became accepted as being no more than equally effective. Some scientists say that the decline effect is simply science self-correcting and working through its errors, although others fear it raises troubling questions about the extent of human biases in the supposedly pure, truthful realm of research.

None of that means that science, technology and human knowledge aren't advancing. They are surging in all directions, but in a far more unexpected and wayward manner than arrow-flight projections predict. As much as I love the regular hit of new concepts and revelations found in the news sections of the scientific press, breaking news stories are not the best guide for predicting the future.

A more useful approach can be found in the work of the Polish author Stanisław Lem, who was active in the second half of the twentieth century. Lem wrote about the internet, virtual reality and the 'alternative facts' post-truth world long before they existed. He was the first to imagine nightmare scenarios such as the technological singularity which so worries AI researchers, or out-of-control nanotechnology turning the world into 'grey goo'. He predicted things like smartphones, ebooks, audiobooks and 3D printing. It was Lem who coined the phrase 'a Theory of Everything', although he meant it as a joke. 'As far as I can tell,' the science writer Simon Ings wrote in the *New Scientist*, 'Lem got everything – everything – right.'

This might give the impression that Lem was some form of technological realist, but that is far from the case. One of his short stories, for example, involves the invention of a machine that can manufacture anything that begins with the letter 'n', which is then asked to create nothing. Lem had no interest in dreaming up improved versions of current technology. 'Meaningful prediction,' he wrote, 'does not lie in serving up the present larded with startling improvements or revelations in lieu of the future.'

The reason why Lem got 'everything right' was because he understood people and he understood history. He personally knew what it was like to be pressed up against the wall by the muzzle of a Nazi machine gun. The future technologies he dreamt up were not the point of his stories, but ways to explore what humanity does and how it acts. He was not the sort of person to get so excited about the idea of a jetpack that he would then assume real people would use them. What Lem teaches us is that if you want to understand how a technology like AI will develop in the future, your best bet is not to listen to the arrow-flight projections of the Church of the Singularity. Your best bet is to understand the people who will be developing, deploying and using the technology.

This is not as easy as it might sound. The generation born in the twenty-first century are already proving to be very different from those who came before.

4.
THE METAMODERN GENERATION

1.

I didn't really understand how different twenty-first-century teenagers are from people raised in the twentieth century until, by happenstance, I found myself in a room watching the 1985 John Hughes teen movie *The Breakfast Club* with a bunch of them.

The Breakfast Club is the story of five American high-schoolers spending a Saturday in detention. Each of the five is a different stereotype – a brain, an athlete, a basket case, a princess and a criminal, as the film describes them. They begin the film with nothing in common, but they bond in mutual opposition to the authority figure, Assistant Principal Vernon.

I was born in 1971. This puts me right in the middle of Generation X and makes me the exact target market for this film. I thought that younger generations would embrace it in the same way my peers and I did, assuming that they weren't too disturbed by the 1980s styling and music. The Millennial generation, who were born between the early 1980s and the mid-1990s, seem to have done this. It is rare to see a mention of this film that does not include the words 'beloved' and 'classic'. But for post-Millennial teenagers, the story is very different.

In April 2018 the *Breakfast Club* actor Molly Ringwald produced a beautifully written article for the *New Yorker* in which she discussed how uncomfortable she had been watching the film with her young daughter. Her focus was on a scene in which the rebellious teen anti-hero character John Bender, played by Judd Nelson, hides from a teacher underneath a table. Bender is shown looking up the skirt of Ringwald's teenage character Claire, and it is implied that he then touches her inappropriately. This is presented as a comic moment.

Writing in the aftermath of the anti-harassment #MeToo and #TimesUp movements, Ringwald's article attempted to understand why this scene was considered acceptable at the time. 'It's hard for me to understand how John [Hughes] was able to write with so much sensitivity, and also have such a glaring blind spot,' Ringwald wrote. She admitted that 'If I sound overly critical, it's only with hindsight. Back then, I was only vaguely aware of how inappropriate much of John's writing was, given my limited experience and what was considered normal at the time. I was well into my thirties before I stopped considering verbally abusive men more interesting than the nice ones.'

From my experience of watching *The Breakfast Club* with post-Millennials, the issue is considerably bigger than this one dodgy scene. I'd go as far as to say that the film no longer makes sense at all to modern teenagers.

Part of the problem is how this generation reacts to the character of Assistant Principal Vernon, played by Paul Gleason. Vernon is aggressive, controlling and condescending. To my generation, he was the antagonist or villain of the story, and it was his cruel exercising of his authority that united the film's heroes. To the current generation, however, Vernon is simply doing his job. They see him not just as a flawed individual, but also part of a necessary system. His character's motivation is simply to improve behaviour at the school. He gains nothing personally from the situation and has even given up his weekend to go into work for this greater good. He's not a likeable character, admittedly, but morally he is not the bad guy. At best, he is light relief.

The willingness of post-Millennials to sympathise with an authority figure and shun the rebel is part of the reason why older journalists have been quick to label them as 'boring'. Their perceived lack of rebelliousness is backed up by numerous studies. Compared to previous generations, today's teenagers are less likely to drink alcohol, go to parties, crash cars, have sex, get arrested or become pregnant. Some journalists have used this as an excuse to call them 'Generation Yawn', 'Generation Sensible' or 'Generation Zzzzz'.

To this generation, the villain of the film is the very character that my generation viewed as the main male hero, the bad-boy teenager Bender. To the eyes of Generation X, Bender was a character who refused to bend to authority. Sure, he was troubled and difficult, but he remained true to himself. He was a self-focused individual who had integrity on his own terms. This was something that we Gen Xers admired.

To the eyes of the post-Millennial generation, Bender is just a bully. He deliberately makes others miserable and takes delight in doing so, and no amount of bad-boy charm or troubled backstory can disguise the fact that the character is an unredeemable arsehole. As Ringwald wrote, 'As I can see now, Bender sexually harasses Claire throughout the film. When he's not sexualizing her, he takes out his rage on her with vicious contempt, calling her "pathetic," mocking her as "Queenie." It's rejection that inspires his vitriol. [...] He never apologizes for any of it, but, nevertheless, he gets the girl in the end.'

Having all this pointed out to me by a modern teenager was quite an eye-opener, because I was unable to argue that they were wrong. Bender is intentionally cruel, and that makes him a horrible person. This leaves us wrestling with the same question as Ringwald – how is it that those of us raised in the 1980s failed to see something that is blindingly obvious to twenty-first-century kids?

Modern teenagers are also not impressed that the alternative loner character, played by Ally Sheedy, is only considered to be desirable after she has been given a more conventionally feminine makeover. In fairness to Generation X, this always seemed wrong, even in the 1980s. What really confuses a modern audience, however, is the film's ending, and in particular the treatment of the nerd character, Brian.

At the end of the film, the four cool teenagers pair off while Brian is left behind to finish the detention essay. This is obviously unfair, but what makes it worse is that Brian confessed to making a suicide attempt in the previous week. To modern eyes, he is a character who clearly needs help and support. Instead, he is placed in detention by the school, told he is a disappointment by his mum, and ditched by the more attractive kids. Today's teenagers find the film's treatment of him to be completely bewildering. Surely Brian's situation should be the dramatic heart of the film?

To my generation, Brian didn't matter. His attempt at suicide, which was botched through use of a flare gun, was played for laughs. It rarely figures in our memories of the film. We are more likely to remember the final shot, in which Bender triumphantly punches the air as he walks away across a football field, than Brian's emotional confession. That post-Millennials pay much greater attention to Brian, and also hold a negative opinion of Bender, suggests that they are considerably more empathetic than their parents or grandparents.

The idea that there's been a generalised increase in empathetic emotional intelligence over the past few decades provides some context for contemporary reactions to historic sexual predators, and long-term lower-level sexual abuse, in Hollywood and other industries. We now understand that 'the times have changed' and 'it was different back then' are not acceptable defences for historic abuse. Yet those are the same arguments that my generation instinctively reach for when we attempt to explain why we overlooked Bender's bullying and abuse, and ignored Brian's suicide attempt, when we enjoyed *The Breakfast Club*. To modern, empathetic teenage eyes, late-twentieth-century people should have known better. And yet we didn't.

2.

For most of history, the move from child to adult was as abrupt and clearly marked as the change from water to ice. Childhood was a time of play, and adulthood was a time of responsibility. Many cultures established traditions to mark this coming of age, which reveal what age the change was historically considered to occur. In Judaism, for example, boys have a bar mitzvah at the age of 13 and girls have a bat mitzvah at the age of 12. By the twentieth century, childhood was usually thought of as lasting a little longer than this, but not much. As late as 1960, the median age at first marriage for American women was 20, meaning that half of American women getting married for the first time were teenagers. If a woman was not married by her mid-twenties, she was in danger of being considered an old maid.

After childhood ended it was time to adopt the responsibilities of adulthood, namely work, marriage and raising children. As St Paul wrote in the Book of Corinthians, 'When I was a child, I spoke as a child, I understood as a child, I thought as a child; but when I became a man, I put away childish things.' But during the 1950s, in the post-war Baby Boomer generation, something unusual happened. A period of adolescence opened up between children and adults. These people weren't children exactly, for they pursued the adult thrills of drink, sex and independence. But neither were they adults, for they did not adopt the responsibility that came with adulthood. A new word had to be coined to describe them: 'teenagers'. That word didn't exist before the 1950s. There had never been a need for it.

After the Baby Boomers came my lot, which the Canadian novelist Douglas Coupland named Generation X. This group is typically defined as being born between the beginning or middle of the 1960s and the beginning of the 1980s. We looked at the Boomers' teenage adolescent period and we liked what we saw. We realised that we could extend it for quite some time – possibly even indefinitely. We were very much in favour of the sex, drugs and rock 'n' roll that flowered during adolescence, we could not see anything of value in the career-minded, responsible adulthood that was supposed to follow. We were called 'slackers', because of our lack of interest in smart clothes and corporate careers. These things struck us as meaningless. We failed to suggest a meaningful alternative, but the questioning of the status quo and the realignment of values which caused it were sufficient to define us.

After Generation X came the Millennials. They were born between the start of the 1980s and the mid-1990s, and are so named because they came of age around the Millennium. Millennials turned away from the irony, nihilism and cynicism of Generation X. They took the extended adolescence that Baby Boomers had begun, and which Generation X had extended, and turned it into something that more resembled a continuation of childhood. Childhood was a place free from Gen-X nihilism and cynicism.

4. THE METAMODERN GENERATION

In previous generations, childhood interests such as superheroes, Disney movies or the Millennials' own *Harry Potter* books would have been dropped in embarrassment when drink and dating entered the picture. Now, they continued to be enjoyed, even by people who were old enough to be raising children themselves. Generation X had begun this trend of infantilization, but we insisted we were only doing so ironically. The Millennials, in contrast, were sincere. It is mainly because of them that the most successful Hollywood output over the past decade has been Marvel superhero films, which are adaptations of stories written in the 1960s and 1970s to entertain eight-year-old boys. To Baby Boomers and many Generation Xers this can seem embarrassing or disappointing, but Millennials feel no shame about it. They see nothing wrong with discussing their thoughts on Iron Man, Captain America and the rest while at work holding down adult jobs.

The non-ironic sincerity of Millennials has been confusing to postmodern Generation Xers, who dismissed sincerity and couldn't grasp the concept of its return, but it has particularly annoyed Baby Boomers. They saw Millennials demanding safe spaces and trigger warnings at universities, or requiring hand-holding and constant praise at work, and wished that this entitled, over-protected generation would just grow up. Of course, Millennials were the product of how the Boomers had raised them. According to the popular stereotype, they grew up being told that they were special by parents, teachers and Barney the Dinosaur. Millennials were given awards just for turning up and congratulated on their participation when they failed exams. They were taught to believe in themselves and told that there was nothing that they could not achieve so long as they remained true to their heart. As a result, they became the most individualistic generation we have ever seen, and can appear self-centred and self-obsessed to everyone else. During their childhood the focus was on them, and they expected to be celebrated, well paid and successful when they left college. The reality of the world of work has been hard on them.

The hate that Boomers project onto Millennials and the hipster culture they invented is quite a thing to behold. To choose just one example, the journalist Will Self wrote a *New Statesman* article in 2014 titled 'The awful cult of the talentless hipster has taken over', in which he bemoaned 'the sheer quantity of dickheads now wandering bemusedly around the world'. This generated a wave of similar articles, including one in the *Guardian* headed 'Is it OK to hate hipsters?'

Baby Boomer and Generation X complaints about entitled Millennials are usually accompanied by a sense of bewilderment. The older generations can't understand why Millennials should care so much about such unimportant matters as male grooming, coffee blends or steampunk cosplay. This is because they do not see sincerely caring about something trivial as worthwhile in itself.

For those raised in the circumambient mythos of the post-modern late-twentieth-century West, there is no meaning to be found in the world. Anyone who thinks otherwise is considered delusional, which is an attitude taken to its extreme by the angry-old-man atheist movement. But to Millennials meaning is something subjective rather than objective. This was how they were able to resurrect the concept after it had been so firmly refuted by the nihilistic Generation X. As the Millennial generation understands the world, if they find something personally meaningful then that means that there is meaning in the world. Surely finding value in obsessive attention to a well-groomed beard was an improvement on Kurt Cobain-style nihilism?

Then, in the mid-to-late 1990s, a new generation arrived.

3.

There have been numerous attempts to name the generation that followed the Millennials, including 'Generation Z', 'Snowflakes', 'post-Millennials', 'Zillennials', 'Digital Natives' and 'iGen'. At the time of writing it is not clear which, if any, of those names will catch on, but Generation Z is currently the most common.

Generation Z started school in the twenty-first century and have never known a world without the internet. They are the first generation to attend high school with a smartphone clutched in their hands. According to the American psychologist and demographic researcher Jean M. Twenge, it is constant access to mobile internet that has defined this group and caused an unheralded shift in generational attitudes. 'Around 2012, I started seeing large, abrupt shifts in teens' behaviours and emotional states,' she has written. 'In all of my analyses of generational data – some of it reaching back to the 1930s – I had never seen anything like it. At first I wondered if these were random blips that would disappear after a year or two. But they didn't – the trends kept going, creating sustained, and often unprecedented, trends. As I dug into the data, a pattern emerged: many of the large changes began around 2011 or 2012.' Twenge examined a host of possible explanations for this sudden dramatic shift. The only one she could find that fitted the data was the large-scale adoption of smartphones at that point.

Post-Millennials are typically not keen on the idea that they have been shaped by smartphones. An anonymous viral meme has spread widely on the internet, detailing the following hypothetical conversation between generations:

Adults: Record numbers of teens are depressed, we must find out why.

Teens: School is more stressful than ever, our parents
    screwed over the economy, the Earth is on a path to total
    environmental destruction and now we have to deal with
    actual fucking Nazis
Adults: It's the iPhones, isn't it?

This is funny, but there is a lot of data to support the idea that
the huge increase in teenage mental-health issues occurred when
smartphones were adopted. For example, according to the US
Department of Health and Human Services, 56 per cent more
teens experienced a major depressive episode in 2015 compared
to 2010. That's a huge change, which other explanations struggle to
account for. Generation Z are spending more time physically alone
in their rooms with their phones. This has a detrimental effect on
mental health, as does the lack of sleep caused by late-night phone
use. Sporadic use of mobile devices is not considered harmful,
and there is much debate about how much use is excessive and
damaging, but there are studies which show that the greater level
of depression in girls than boys is linked to time spent on social
media, and that unhappiness and mental-health issues are linked
to use that exceeds two hours a day. Most Generation Z phone use
is way beyond that mark.

Older generations lived through times when global thermo-
nuclear war seemed possible and, at times, probable, but even
issues like this didn't produce the same increase in anxiety and
depression that Generation Z are experiencing. The concerns that
the fictional teens in the above spoof conversation have are real
issues, but it is the constant chirping of the phones in their hands
that is informing them about those issues.

The skyrocketing levels of teenage anxiety have led to the
anti-anxiety prescription medicine Xanax becoming a much-in-
demand street drug. It has spawned its own subculture and has

been glorified by young rappers such as Lil Xan, most notably on his album *Total Xanarchy*. The Xanax-fetishising rap scene is easily recognised by its slurred, mumbled lyrics and its fondness for ill-judged facial tattoos. The dangers of the drug became clear after the rapper Lil Peep posted an Instagram video where he appeared incoherent and confused and was having difficulty getting Xanax pills into his mouth. He was found dead shortly afterwards of an overdose of Xanax and other drugs.

Just as Baby Boomers are quick to criticise Millennials, Generation X seem particularly annoyed by this overly anxious post-Millennial generation. In the few years since Generation Z started college, Generation X journalists have written many scathing articles about their desire for safe spaces, trigger warnings and emotional security. These trends began in the late Millennial era, but they have been enthusiastically embraced and championed by Generation Z. To Generation Xers, who grew up in a world shaped by the Cold War, such a fearful and mollycoddled generation is baffling. It does not help that Generation Z are so strangely contradictory, being both extremely liberal and extremely conservative at the same time. They fiercely reject any inequality based on race, gender or sexual orientation, but they respect authority, are obsessed with safety and fear economic uncertainty. They show great empathy for others but suffer from social anxiety to the extent that they will go to great lengths to avoid face-to-face communication with strangers. They are less rebellious but also increasingly unprepared for responsibility. They have taken the extended childhood that the Baby Boomers began, and which the Millennials perfected, and turned it into a strange mutation of adolescence that is neither childhood nor adulthood, but which oscillates between the two positions as circumstances dictate.

4.

When I describe demographic groups such as Baby Boomers or Millennials, I am of course generalising.

Generalising can make us suspicious, especially when we have personal experience that seems to contradict such sweeping statements. It is all very well claiming that today's teenagers suffer from social anxiety and demand safe spaces, you might think, but how can this be true given the violence between inner-city gangs at gigs for hardcore forms of rap, such as drill, that you read about in the newspaper? Or perhaps you find the statement that Generation Z are supportive of LGBT issues doubtful, because you recently heard a teenage member of your extended family making homophobic jokes?

Psychologically, we're still not very good at grasping the difference between the world we have experience of and the larger-scale patterns we find in big data. This is why people, including powerful elected officials, can argue that climate change can't be real because it snowed at Easter. In an attempt to counteract this, environmental scientists repeatedly explain that there is a big difference between climate and weather.

In a similar way, broad demographic categories such as Baby Boomers and Millennials need to be thought of as more like climate than weather. The individual data points are not what is important. The patterns that become visible when huge quantities of individual bits of data are analysed are what is important. As meteorologists will tell you, the huge, inhuman-scale overview that we call climate is far more useful and important than any individual record of rainfall.

The claims I was making about how past generations differ are based on the work of demographic researchers, and in particular the books of the American psychologist Jean M. Twenge. Much of Twenge's work is based on four publicly available American data-bases. These are the General Social Survey, which has examined adults since 1972, Monitoring the Future, which asks schoolchildren thousands of questions and has run since 1976, the Youth Risk Behavior Surveillance System, which has run since 1991, and the American Freshman survey, which began in 1966. All told, these surveys have questioned 11 million people and are representative of the US population in terms of race, gender and socioeconomic status. Importantly, such long-running surveys allow us to com-pare attitudes of children at the same age over time, rather than being a snapshot of how the young and old differ at one particular moment. The focus on America is necessary because few countries have such high-quality data, but the conclusions drawn from the US do seem to be broadly in line with other Western countries. For example, British research tells us that a third of 16- to 24-year-olds in 2015 don't drink alcohol, compared with around one in five 10 years earlier, and this change matches a similar American decline reported by Twenge. In a similar way, the data we have from countries including Sweden, Holland, the UK, Australia, Finland and Japan all points to young people having less sex than before, just like in America.

It is easy to zone out when confronted with dry statistics. It is more human to talk of individual experience and anecdotes even when these may not be typical. My account of watching *The Breakfast Club* with members of Generation Z is an example of such an anecdote. I include it here because I feel comfortable that it matches and illustrates the attitude changes found in generational studies. Politicians are also often wise to the fact that an individual story has more impact than the results of large-scale data analysis, although they are not always so careful to choose a story that is representative.

As useful as big data is, there is always an arbitrary nature about how categories are defined. Earlier, I discussed those born after the Second World War in terms of four generations – Baby Boomers (those born between 1946 and 1964), Generation X (1961–1981), Millennials (1982–1996) and post-Millennials (1995–2012). Or, at least, those are the dates for those generational categories if you ask Google. When you examine individual generational studies, it quickly becomes apparent that different researchers favour different dates. Different definitions of Baby Boomers, for example, claim they begin either in 1940 or in 1946, and end either in 1960 or in 1964. There is no 'official' definition of those categories. Someone born in 1962 can be called a Baby Boomer or a Gen Xer with equal validity. We think that data is precise, but it is a tiny snapshot of the undulating mess that is the real world.

Those four named generations are a decades-long outpouring of millions of newborn children reduced down into four arbitrary data points. Those data points suggest that a Generation Xer born on New Year's Eve 1981 is widely different to a Millennial born just a few hours later, on New Year's Day 1982. They would also imply that our 1981-born Generation Xer would be the same as someone born 20 years earlier, in 1961. This is clearly not the case. And yet, looking at generational changes through the flawed model of these four categories teaches us more than if we did not use this model. The generational categories may be vague and too loosely defined, but they do reveal distinct changes across time, and this stops us from making the all-too-common mistake of assuming that other generations think like we do. They also produce patterns, which we would otherwise be blind to. Here again is the contradiction at the heart of the big data industry. With data, quantity is as important as quality. Data may be flawed, but it can still be insightful. Even with these problems, we can still learn a lot from it.

5.

In biological terms, humans had remarkably lengthy childhoods long before teenagers were named. Chimps and gorillas are considered adults, and are ready to reproduce, at the ages of six and eight respectively. To find non-human animals that take until their teens to become sexually mature you need to look to mammals as large as elephants and whales. Looking back in time, we find that our australopithecine ancestors had childhoods of a similar length to apes until around a million and a half years ago. At this point, the biological childhoods of our ancestors began to lengthen. This process continued until around 100,000 years ago, after modern man had appeared in Africa.

Long childhoods have many disadvantages in evolutionary terms. They slow down the rate of reproduction and create a longer period of dependence during which the child needs to be fed and protected. While our early *Homo sapiens* ancestors had a long biological childhood that was similar to our own, our close cousins the Neanderthals had much shorter childhoods. That we succeeded over the Neanderthals suggest that long childhoods might offer some evolutionary advantage that is sufficiently important to outweigh the obvious disadvantages of long childhoods. This does appear to be the case: longer childhoods produce more intelligent adults.

This is not simply a question of brain size, because our brains stop growing before we become adults. Instead, it is a question of how those brains are wired up. Our lengthy childhoods allow us time to develop social intelligence, creativity and the ability to collaborate with our tribe in increasingly elaborate ways. As the American anthropologist Agustín Fuentes explains, 'Whether it was eluding predators, controlling fire, telling stories, or contending

133

with the shifts in climate, our ancestors creatively collaborated to deal with the challenges the world threw at them. At first they did so in ways that were just marginally more effective than those of their pre-human forebears and other humanlike species. Over time, that minor edge of advantage expanded, refined, and propelled them into a category all their own.'

It is not the case, current evolutionary thinking tells us, that it was the competitiveness of cruel and cunning alpha males which set us apart from the rest of the animal kingdom. On the contrary, it was our ability to share and co-operate to a greater degree than other species. This focus on community is a central aspect in the emergence of language, culture and technology. As Fuentes describes our evolution, it is 'the story of a group of highly vulnerable creatures – the favoured prey of a terrifying array of ferocious predators – who learn better than any of their primate relatives to apply their ingenuity to devising ways of working together to survive; to invest their world with meaning and their lives with hope; to reshape their world, thereby reshaping themselves.'

It is partly through play that we learn social intelligence and creativity. Much of this takes the form of imaginative make-believe games, which allow children to explore imagined scenarios in a safe, curious way. The longer children spend doing this, the more prepared they become for dealing with the unexpected situations they encounter as adults. This idea is becoming increasingly accepted in early education and is nicely captured by the title of a 2004 book by the American academic Roberta Michnick Golinkoff, *Einstein Never Used Flashcards: How Our Children Really Learn – and Why They Need to Play More and Memorize Less.* There is evidence for this approach in the success of early education in countries such as Finland and Sweden, which prioritise play-based learning over formal education until the age of seven. These children quickly catch up with, and in many cases overtake, the academic achievements of children in countries which begin formal teaching at the age of five.

The cultural lengthening of childhood over the past 100 years was not the result of biological evolution, because evolution doesn't work over such short time periods. If anything, improvements in diet and healthcare caused physical puberty to occur earlier than before. The changes had social and cultural causes, rather than biological ones. Further carefree play, rather than settling down to responsibility, was necessary to make us capable of coping with the constant changes that the twentieth century was throwing at us. It seems that this approach worked, as average IQ scores increased by 30 points over the past 100 years, in a phenomenon known as the Flynn effect. Intriguingly, a 2018 Danish study suggests that the Flynn effect went into reverse during the last quarter of the twentieth century. Potential explanations for this reversal have varied, but some claim that the nature of intelligence itself has changed in a way that the standardised IQ tests fail to measure. This is something we will return to.

A key cultural change was the explosion of media over the last century. This caused an unprecedented increase in the amount of information about the world that our brains had to deal with. For most of history, the majority of what we knew of the world came from immediate first-hand experience of the places where we lived. This could be supplemented by the telling of stories, singing of folk songs or performances by travelling actors, but the majority of what our brains had to process concerned our immediate environment and experiences. The invention of the printing press changed things radically, but books were initially prohibitively expensive for most people. Printed pamphlets followed, which were cheaper and more widespread, but only for those who could read. By the start of the twentieth century, not only had reading become common but cinema and radio arrived. Our brains began to process the sights and sounds of the world beyond our own horizons. For the first time we knew what countries on the far side of the world looked like without having to spend years travelling to find out.

The media-saturated twentieth century was only getting started. Soon there was black-and-white TV, and the affordable 7-inch single. Colour TV came along, quickly followed by video cassette recorders, an explosion in the number of channels, and the fast-cutting of images pioneered by the music station MTV. Early arcade games arrived, followed by home computers and modems. The 1980s and 1990s gave us fax machines, CD-ROMs, Nintendo consoles and the fledgling World Wide Web. By the time Apple launched the iPhone in 2007 and constant mobile internet became an ingrained part of our daily lives, our brains were processing far more information than they had a century before. Making playful childhood longer, to make us smart enough to navigate the hyper-mediated modern world, not only made sense but was entirely necessary.

Another important change was life expectancy. We are now living roughly twice as long on average than we were a century or so ago. In 1900, the average life expectancy at birth in England and Wales was 48 for women and 44 for men. By 2010, that had increased to 89 for women and 85 for men, and the world average for both sexes shows a similar leap, going from 39 in 1900 to 71.5 in 2014. When you expect to live longer, there is less pressure to settle down and raise children quite so quickly.

Historically, we had evolved in cultures where the world was more or less the same over generations. In circumstances like these, a brain formed by infant play could stop evolving at the end of childhood and still understand the world its adult self would encounter. But thanks to the unprecedented rate of technological change in the twentieth century, a person who grew up with horse-drawn transport could live to watch a man walk on the moon. A brain shaped by childhood play in the early twentieth century would have to make sense of the radically different late twentieth century, especially now it would typically live far longer than the historical average. Our current difficulty with grasping the insights of big data compared to individual experience is a good example of this, because we were not taught about big data at school. A brain designed to understand the world as it was during childhood, rather than the world adults find themselves living in, is going to encounter problems later regardless of how intelligent it is.

A 'generation gap' first appeared in the mid-to-late twentieth century, because older people found themselves unable to understand the culture of the young. This is also why a journalist as bright as Will Self can be so incapable of understanding Millennial hipsters that the only conclusion imaginable to him was that they are all dickheads.

When seen in this light, the fact that childhood has been culturally extended over the second half of the twentieth century makes a great deal of sense. Remaining childlike and playful was the obvious adaptation we needed to make. Creative play is still one of the best ways to keep learning and to becoming comfortable with new perspectives. People need a safe way to mess around with new technology and new ideas. That Millennials and post-Millennials are taking longer to grow up and seem infantilised to Boomers is, when seen from this perspective, entirely to their credit.

6.

The moment when I realised that younger people thought differently to me came after I read articles about the behaviour of students in American universities. One of the most influential of these was a September 2015 cover article in the *Atlantic* by Greg Lukianoff and Jonathan Haidt entitled 'The Coddling of the American Mind'. It was illustrated with a photograph of a bewildered-looking curly-haired toddler, who was sat at a desk with a laptop, wearing a sweatshirt that read 'College'. The article's strapline read, 'In the name of emotional well-being, college students are increasingly demanding protection from words and ideas they don't like. Here's why that's disastrous for education—and mental health.'

For many Generation Xers like me, this article was an introduction to Millennial and post-Millennial concepts such as trigger warnings, microaggressions and safe spaces. University students had started to demand the right not to encounter ideas they found unpleasant. Lukianoff and Haidt's essay talked about how Harvard law students had asked professors not to teach rape law in case it distressed the students. In one case, professors had even been asked not to use the word 'violate' in sentences such as 'violates the law'.

To my generation, examples like these sounded ludicrous. As we saw it, law students are training to be lawyers, and they will encounter things far more distressing than words once they begin work. If universities provided them with 'safe spaces' where they were guaranteed not to be exposed to ideas that they found offensive, then they were going to be in for a hell of a shock after graduation when they would be spat out into the world with no institutions to protect them.

The temptation was to assume that anecdotes such as these were extreme examples, perhaps exaggerated through biased retelling. It's increasingly clear that this isn't the case. Jean Twenge surveyed hundreds of Generation Z students, aged 21 and under, and concluded, 'these are not fringe ideas but those embraced by the majority [of that generation]'. Countless professors and researchers have reported similar attitudes across the academic world, much to the delight of newspaper editors.

Thanks to our lengthening childhoods, university students should now be thought of as older children, rather than the young adults of previous generations. Their extended childhood means that they are not always ready to stand on their own two feet. Many are expecting universities to offer a similar protective role to that provided by their parents at home.

What is curious about the notion of safe spaces is that this generation is, physically, the safest there has ever been. Most physical risks, from traffic accidents to physical assaults or passive smoking, have dropped dramatically during previous decades. As children, this generation were less likely to roam free in their neighbourhoods, getting into scrapes and learning about the consequences of bad ideas the hard way. Instead, their parents often arranged playdates for them and scheduled their activities. They were constantly monitored to the extent that the classic parenting phrase 'they're around here somewhere' has largely fallen out of use.

Teenagers' concerns about safety are not about physical injury but instead about *emotional injury* – which in itself is a phrase almost entirely confined to the post-Millennial generation. As Twenge writes, 'Perhaps because they are so physically safe compared to previous generations, and perhaps because they spend so much time online, [Generation Z] sees speech as the venue where danger lies.' She goes on to quote a 20-year-old boy who tells her, 'There is physical danger and emotional danger. Traumatic experiences can affect your mind and cause emotional suffering which can feel just as negative as physical suffering.'

This is, I think, the thing that my generation have difficulty grasping. When we hear of students demanding safe spaces to protect themselves from ideas, disinviting speakers with whom they disagree, or attempting to have courses changed and professors sacked, we can't comprehend why ideas threaten them to such an extent that they feel compelled to avoid them. Unconsciously, we agree with the old adage that sticks and stones may break my bones, but words will never hurt me. We give our internal mental landscape less weight than that of the external world.

What has occurred in more recent generations is an elevation of the significance of the internal world. What this generation think and feel has become as important as – or more important than – events in the material world. This change helps explain the decline in the amount of teenage drinking, sexual activity and drug use. It used to be that teenagers were drawn to these activities because of the physical pleasure they provided. But internal emotional needs have now gained primacy over sensations and material wants, so these behaviours are becoming less common.

For this generation, their anxiety comes not from physical dangers, but from the risk that they may encounter something that hurts them emotionally. From this perspective, every conversation is a risk. The fact that they prefer to queue for an automated till in a supermarket than talk to a cashier they do not know is a consequence of this. As the polling and research company YouGov found in December 2017, when it examined the willingness of British people to engage in conversation with strangers in a range of scenarios, 'A key trend revealed by the results is the reluctance of Millennials to talk to other people. Across all nine scenarios, Brits aged 18–34 are noticeably more likely to say that they would prefer not to talk, with the biggest generational gaps [in attitudes to talking to people] on public transport and tradespeople.'

Recall how this generation has reacted to advances in AI and robotics. While Baby Boomers and Generation X often treat them with suspicion, Millennials and Generation Z appear drawn to them. The demographic differences in attitudes to automated supermarket tills is a clear example of this. It also explains the effort Google put into training AI to book appointments over the phone. One factor in the uptake of AI and robots, I suspect, will be the upcoming generation feeling more comfortable with digital interactions than conversing with strangers. They know that a machine will not be deliberately hurtful.

This concern about emotional injury has occurred in a generation largely kept safe in their bedrooms, conversing with their friends through the filter of mobile internet. Compared to face-to-face dialogue, online communication frequently lacks empathy. For example, if you have a face-to-face conversation with someone and they say something that you don't entirely agree with, the natural response is to let it slide. Perhaps it was just a passing thought, verbalised without a great deal of consideration. It may have been irrelevant to the larger discussion, or a perspective that is plausible as part of a larger argument. Even if it was something that was just flat-out foolish, there is still little reason to raise the matter and get into an argument. Yet when you see the exact same words appear as text on social media, your response can be very different. Those words might induce anger, or a sense of appalled self-righteousness. You may feel compelled to publicly condemn the speaker, perhaps dramatically declaring them dead to you. The lack of empathy in online communication is particularly evident on social media.

To compensate for the lack of online empathy, this generation has pioneered communicating with emojis, and with pop culture gifs in which a character expresses a particular emotion, because they recognise that online communication does not come with the same body language that accompanies face-to-face conversations. This lack of body language is particularly problematic for brief text exchanges which don't provide a great deal of context for interpreting the intended tone of the message. You can see how a generation raised online would be grateful for the emoji language which consists of nothing but visual symbols for nouns and emotions.

Perhaps it is because of the lack of body language online that post-Millennials have become extremely skilled at inferring the thoughts and feelings of others. They repeatedly show more understanding of group dynamics than their older relations. One example of this is their reaction to *The Breakfast Club*, as mentioned earlier. Another is a popular meme based on the 1991 Disney movie *Beauty and the Beast*. This shows the film's heroine, Belle, singing the film's opening number, during which she wanders through her home village commenting on its inhabitants. The first image shows Belle singing the line, 'There goes the baker with his bread, like always'. This is followed by a picture of the baker in which the meme maker has given him the line, 'There goes Belle with her daily mean song about us'. The joke recognises that Belle had been going around her village insulting or dismissing people, which is not a nice thing to do. This is something that never occurred to the late-Generation Xers and early Millennials who queued up to see the film on its release. It is not as if no one gave any thought to the song. It received a Best Original Song nomination at the Academy Awards and has been the subject of a great deal of analysis by film critics. The *New York Times* film critic Janet Maslin, for example, wrote that 'This rousing number reaches such a flurry of musical counterpoint that it recalls sources as unlikely as *West Side Story*, while the direction builds energetically from quiet beginnings to a formidable finale.' The song was thoroughly analysed and discussed. Yet it took the young to see the song not only from Belle's point of view, but from the point of view of the other villagers as well.

The emergence of the phrase 'my friendship group', and its colloquial cousin 'the squad', is another example of this shift. I've never heard anyone in Generation X or older refer to their 'friendship group', although its origins possibly lie in the increased popularity of the phrase 'circle of friends' during the 1990s. Before then, people would just say 'my friends'. The focus of 'my friends' is on the self and its relationships; at the centre of this world there is you and connected to you are your friends. In contrast, the phrase 'my friendship group' takes the focus away from you and focuses on the wider group of connected people, of which you are just one. It understands the shifts in dynamics when a new person is admitted into the group on a deeper level than just your personal connection to that person. The phrase recognises relationships between your friends which don't necessarily involve you.

The arrival of this wider viewpoint in the younger generation has helped create an emotionally wiser and more empathic culture. The original audience for programmes such as *The X Factor* used to love seeing untalented contestants humiliated and laughed at, by both the judges and the studio audiences. These would typically make up the 'water cooler moments' that everyone talked about after the show, and the online clips that producers hoped would go viral. Now, producers of shows like this avoid cruelty for its own sake and focus more on heart-warming moments of genuine human connection. This is not because TV producers have become better people, but because the new audience they are trying to reach have different values.

In the charity single 'Do They Know It's Christmas?' by the supergroup Band Aid, the Irish singer Bono sang the lyric 'Well tonight thank God it's them instead of you' in both the original 1984 version and the 2004 re-recording by Band Aid 20 (the same line was sung by Jason Donovan and Matt Goss in the 1989 Band Aid II version). Bono returned again in 2014 for Band Aid 30, but by this point generational changes had made the line feel too brutal and upsetting. Bono's lyric was neutered to become, 'Well tonight we're reaching out and touching you'. It's not as good a lyric, if we're being honest, but it is more in tune with the new teenage audience.

A clear indicator of the shift in emotional intelligence between the Millennial and post-Millennial generations can be found in their attitude to sex. Millennials were the most individualistic of all generations, and their attitude to sex focused primarily on self-gratification. Their 'hooking up' culture of casual sex attempted to separate emotional bonding and human connection from sexuality. To the post-Millennials, and indeed to most other generations, this is to miss the point quite spectacularly. Post-Millennials coined phrases like 'all the feels' or 'right in the feels' because they are deeply focused on the emotional side of life.

It is easy for my generation to scoff at all this, but it has long been argued that many of the problems of the twentieth century were caused by it being too cold, analytical and patriarchal, and that a shift to a more female, emotionally intelligent culture was desperately needed to counter its excesses. Looking at the post-Millennials, it seems that this is exactly what is happening. As noted earlier, IQ scores rose over the course of the twentieth century in a phenomenon called the Flynn effect, with a 2015 study finding an average increase of 20 points since 1950 over 48 countries. There is now concern that rises like this appear to be coming to an end not because people are becoming more stupid, but because our current IQ tests are not designed to recognise a shift towards emotional intelligence over intellectual intelligence.

A small but telling example of this change could be seen in the 2018 Brit Awards, when Dua Lipa won the award for the Best British Breakthrough Act. Lipa was born in 1995, in the grey area between the end of the Millennials and the start of Generation Z. When she went up on stage to collect her award, she took her younger brother and sister up with her because she wanted her loved ones to feel part of the moment. Out of the hundreds of winners in the 40-year history of the awards, this was not something that anyone else had ever done. It was a simple, casual act in many ways, but it beautifully captures the different priorities that mark how Generation Z differ from their elders.

This new emotional and empathic awareness, we should note, is coming from a generation that is still keeping one foot in childhood. The emotions being recognised are largely the emotions of childhood. Mature, nuanced examples of emotional intelligence are noticeable by their absence. You will be hard pressed to find examples of forgiveness in the world view of the young, for instance.

7.

If we want to understand how this increase in empathy arose, it can help to look at the broader cultural changes that accompanied it.

Until the arrival of the internet, Western thought in the twentieth century focused on the primacy of the individual. The internet brought with it the realisation that the individual is too simplistic a model to explain society, and that the network is a more powerful and insightful concept with which to understand our lives. Generation Z followed the highly individual Millennials but grew up in a world of constant mobile internet connections, so they simultaneously saw themselves as proud individuals and also as networked members of larger, fluid groups.

Generation Z inherited from the Millennials a belief in the importance of individuality and a belief that they had the right to be exactly who they wanted to be. But their ability to think in terms of networks added an extra insight to this, which is that individuality can only work when *everyone* is allowed to define who they are. It is only when everyone accepts how everyone else defines themselves that the world can be trusted to allow you to be yourself. A vital part of the right to individuality, therefore, is to respect how others see themselves.

There is a cost to this. It requires an increase in empathy in order to constantly assess how others are feeling and how they see themselves. It is a lot of work for the brain to process and keep track of a group of wildly different, unpredictable individuals. It is far easier to grasp a group of identically dressed interchangeable clones who exhibit predictable behaviour. Remembering the preferred pronouns of one or two people is easy, for example, but it becomes more difficult when there are a dozen or more. Yet the fact that the young keep celebrating diversity suggests that the effort is worth it. You are never truly free to be yourself unless you accept others for who they want to be also.

For much of Generation Z, the idea of prejudice against people based on sexuality or ethnicity is so self-evidently absurd that they can sometimes be blind to the institutionalised bias in society. They can dismiss initiatives like positive discrimination, for example, because they can't believe that anyone would use race as a factor in hiring. You can find examples of Baby Boomer and Generation X men and even women who are not happy that Doctor Who has become a woman, but good luck trying to find anyone born this century who finds this in any way controversial.

One subject that shows how Generation Z's acceptance of others is changing wider society is transgender awareness. This developed so suddenly that it took many people unawares, including some who now identify as trans. Previous social movements such as gay rights, civil rights or universal suffrage were marked by decades of dedicated campaigning, which gradually raised awareness in society and the legal system. This did not happen with trans rights. It was a subject that Generation Z immediately understood, and then explained to the rest of us. And when they did this, a great many things that had seemed odd suddenly started to make sense.

Growing up, I was a big fan of Prince. I saw him live during his eighties peak and considered 'If I Was Your Girlfriend' to be the best track on *Sign o' the Times*. I listened to him singing as his female alter ego Camille and watched as, while always remaining clearly heterosexual, he changed his name to a combination of the symbols for male and female. I had no idea what any of that meant because I did not know that gender and sex were different things. I simply assumed that if Michael Jackson could sleep in oxygen tents and have a monkey for a best friend, then Prince could be as weird as he wanted also. Becoming a megastar did seem to make people unusual. What Prince was actually trying to express went completely over my head.

Nowadays, I understand that sex is biological and usually binary, with people who possess XY chromosomes being male and people with XX chromosomes being female. Gender, in comparison, is a spectrum of masculine and feminine traits, and the connection between biological sex and personal gender is largely cultural. Before the arrival of Christianity, for example, Native American people recognised five genders. Sexual preference, of course, is a third category that may be influenced by sex and gender but is ultimately independent. I know all this now because my teenage children explained it to me. I realise that many Baby Boomers deny this and insist that sex and gender are not separate things, but these Boomers are unable to make sense of my Prince records, so I think that Generation Z have a better handle on this one.

Wanting to understand this sudden change in awareness, I go for a drink with the stand-up comedian, musician and author of *A History of Heavy Metal*, Andrew O'Neill. If anyone can shed any light on how the change in trans awareness came about, it is probably Andrew. I find him at the bar in a newly opened pub on Brighton seafront. He is wearing leather trousers with a long flowing black garment that is either an odd-shaped dress or a strangely lengthy shirt. He has long black hair and while his make-up is not excessive, it helps to give his eyes a look of giddy excitement. He looks not unlike how Phoebe Waller-Bridge might look, had she been accidentally cast in the role of Ozzy Osbourne in a Black Sabbath biopic.

Andrew is attempting to buy a beer. This is proving difficult, for he is vegan. Because this is a newly opened Brighton pub, it has an extensive range of obscure craft beers, none of which either of us has encountered before. He asks the barmaid which of the ales is suitable for vegans, and she replies that it is her first day in the job and she has no idea. Unfazed by this, Andrew pulls out his phone and attempts to find information on obscure vegan craft ales. The signal in this underground bar is spotty and this takes some time, but Andrew is not discouraged. Long after I would have given up, apologised for taking up so much of the barmaid's time and ordered a glass of water, he continues his search for information. This, I realise, is Andrew in a nutshell. He is technically a member of Generation X, but his certainty about his identity makes him seem more like a Millennial who arrived early, before the world was ready. He is entirely unapologetic about who he is. He is a vegan transvestite heavily tattooed occultist whose hobbies include ritual magic and camping without a tent and he is not going to pretend otherwise.

When we finally get our drinks, I ask Andrew how the public reaction to his appearance has changed in recent years.

'The difference has been radical,' he says. 'I don't get shit on the street any more, really, for cross-dressing. I first came out as a transvestite when I was nineteen, and that would have been '98 or '99. Even my friends were uncomfortable with it, and they were punks and people who were otherwise right on. I have a distinct memory of my twenty-second birthday, walking from my flat on Camden Road down into Camden and being told by two women on the same road crossing, but out of earshot of each other, to my face, that I was disgusting. It was just like – wow. It happened once, and then it happened again seconds later, the same phrase, "You're disgusting". It was really strange. It ruined my day.

'When groups of guys say anything, what they do is go, "Whoa I thought that was a bird!" or "That's a bloke!" It's technically not an insult. It's kind of like they're just figuring it out, rather than anything more aggressive. There have been some exceptions. But it used to be fairly relentless and now it's really not. It might be due to an increased consciousness of gender fluidity, or it might just be that I'm doing it better. On the whole, on the street, it's really not an issue any more.'

He takes a swig of his beer and nods approvingly in a blokey manner. His choice was worth the effort.

'I have gender dysphoria and it waxes and wanes,' he tells me. Gender dysphoria, as defined by the NHS, is a condition where a person experiences discomfort or distress because of a mismatch between their biological sex and their gender identity. 'Over the years I've worked through it and have reached the point where if I successfully express and present myself in a feminine way, that is my way of treating my dysphoria. I identify as male partly because I have a masculine body and I have a masculine voice. I never, ever get read as female by anyone who's actually interacted with me, even if I lighten my voice and I've done the make-up really well. If people ask which pronouns I prefer, it's basically "he" and "him"unless you're feeling generous.

'I refer to myself as a transvestite as that just seems to me to be the most comfortable fit. I know the word is falling out of favour and it can be seen as sort of seedy and secretive, the shame of middle-class 1970s bank managers, but I'm hanging on to it. There's a big argument about "tranny" being a slur against trans women, but I feel awkward about that as what they are basically saying is "You know the thing that you are? That's a slur, that is."'

'When did things change?' I ask.

'It happened really quickly. I would say it was about five or six years ago when it really started to feel significantly different.'

This would put the change in attitudes, as experienced by Andrew, around 2011–12. This is the exact same time that Jean Twenge started to notice radical shifts in the data about demographic attitudes and behaviours. This is also the same period when mass take-up of smartphone use arrived, causing teenagers to emotionally develop alongside personal, constant mobile internet.

'It was a broader change than just gender' says Andrew. 'It was an increase in understanding, for example about mental-health issues. Remember how comedians like Jasper Carrott and Eddie Izzard used to do jokes about the nutter on the bus? And everyone would laugh at the idea that there is always a nutter on the bus and how you had to out-nutter them. That just wouldn't work any more. The audience would understand that the person had Asperger's and would feel sympathy for them.

'The increased awareness of gender I found really interesting because of how it impacted on my stand-up. When I first started cross-dressing on stage – you're talking maybe fourteen years ago – the difference between audience attitudes then and now is just black and white. It's incredible. The feeling I got was that suddenly the material I had about cross-dressing wasn't working as well. I noticed it first in student gigs. My opening line for years, when I'm cross-dressing on stage, is to do a bit and then point to an alpha male in the front row and say, "What the fuck are you wearing, weirdo?" In a boisterous weekend club gig, that's a good way to acknowledge the elephant in the room. I show the audience that I'm aware of it and I'm being quite alpha about it. That worked less and less over the years. I still do that line sometimes, but at a student gig that just won't work any more. It's just not a thing. The audience, instead of going, "Oh, he's being funny about this thing we think is ridiculous", I would find them going, "Well, we don't have a problem with it, what does it matter?"

'Most of my old material about cross-dressing worked on the assumption that the audience essentially disapproves of it or thinks it's ridiculous or whatever. Now, I have to invoke the disapproval of a third party, in order to get the same jokes to land. There are regional differences, of course. I did a gig in Llanelli the other day, to an older crowd...'

He goes quiet for a moment, with a haunted look on his face.

'There's a difference between here and Australia,' he eventually continues, 'which is the other country I've gigged in the most. Oz is still pretty transphobic. But nowadays, in much of England, I can pretty much get away with cross-dressing on stage without mentioning it. You do get something of a free pass if you look alternative. Interestingly, in terms of relatability with my stand-up, I find audiences get on board with me easier if I'm cross-dressed than if I'm not, because if I'm not, I'm generally dressed like a heavy-metal fan. That's stranger and weirder to most people these days because the old music subcultures are not really a thing any more.

'It's taken me a long time, but I've got to where I need to be. I'm myself now. It used to be that the issue of whether I would cross-dress on any particular day was a big deal. I used to write a blog – which is still up, at genderspanner.wordpress.com – where I would do things like go out cross-dressed every day for twenty-eight days and record what impact that had on me. I stopped writing that blog in 2013, after the great sea change, because I didn't need to do it any more. Dressing doesn't feel like a binary any more. It used to be either I'm going to cross-dress or I'm not, and now it's much more fluid and much more of a spectrum rather than a binary. I've found a coherent self-image that encapsulates both sides of my gender presentation. I basically just dress as myself now and the world has changed to accept me. Or, at the very least, it's stopped calling me disgusting to my face, which is much appreciated. Thanks, world.'

A few weeks after this conversation, Andrew grew a straggly moustache, of the type favoured by early American thrash-metal bands. That quite suited him as well.

8.

Cultural changes occur on different timescales. While looking at things at the scale of generations is useful, it can also be helpful to pull back and take a longer perspective.

In the years after 1900, our sense of certainty took a battering from seemingly every direction. We began that century confident that our understanding of subjects as diverse as space, time, empire, mathematics, the mind, art, language and the material world was solid and trustworthy. Then along came Einstein, Picasso, Freud, Planck, Gödel, Korzybski, Lorenz, Leary, Wittgenstein and a host of others, who introduced us to quantum mechanics, relativity, the unconscious, chaos mathematics, cubism, psychedelics and all sorts of bewildering, anti-intuitive discoveries about the nature of things. They left us without any universally recognised absolutes to orientate ourselves by, and as a result our culture evolved over the course of the twentieth century to accept this new perspective. The changes that appeared in our culture as we attempted to absorb our new understanding of ourselves, when they are grouped together, are what we have named 'postmodernism'.*

---

* There's a lengthier discussion of these matters in my book *Stranger Than We Can Imagine: Making Sense of the Twentieth Century*.

4. THE METAMODERN GENERATION

Postmodernism can be a tricky word to define, because if you ask philosophers what it means you typically get a different answer than you do when you ask artists or social scientists. In general, the defining characteristics of postmodernism include throwing together disparate elements that shouldn't logically fit together, a refusal to be categorised as 'highbrow' or 'lowbrow', a sense of self-awareness that is not afraid to comment on itself, a denial of authorial intent and a willingness to be interpreted in many different ways, and an almost transgressive glee in the act of being postmodern. This may seem like a strange jumble of trends, but that was the logical end product of all that we had learnt in that complicated century.

Although the academic world had an obsessive fling with post-modernism in the post-war years, that relationship quickly soured, and now many intellectuals, particularly those in the Baby Boomer generation, regard postmodernism as a terrible mistake. Writers and thinkers as different as Steven Pinker and Jordan Peterson will tell you it is the cause of all our ills and a misstep that we need to go back and correct. The public had less of a problem with it, however. Outside of the academic world postmodernism was evident throughout mainstream late-twentieth-century culture, as visible in videogames as it was in comedy programmes, dance music and even the outfits worn by Doctor Who.

Postmodernism denies there is an external absolute by which things can be judged. Academia views itself as an external system by which all things can be judged, so it is not surprising that many academics see postmodernism as the bogeyman. It is often por-trayed as the belief that everything is relative, there are no truths and all beliefs are equally valid. This is something of a strawman version of postmodernism, and you would be hard pushed to find an actual postmodernist for whom that accurately represented their beliefs. A more common understanding is nicely articulated by the philosopher Chris Bateman, who wrote that 'The possibility of different practices giving varying perspectives on what is true does not undermine anything but the most simplistic concepts of truth. What is *true* is the sum of everything that can be reliably witnessed: that all these diverse accounts require reconciling is not the same as saying that nothing is true.' That said, the twentieth century also gave us the philosopher Wittgenstein, who argued that the meaning of most words is defined by how they are commonly used, so perhaps we should accept the strawman definition of postmodernism here.

This absence of absolute truth was absorbed into the defining circumambient mythos of the late twentieth century. This post-modern world view peaked during Generation X, and it explains a lot about the nihilistic slacker culture that followed. What, though, has replaced it? If the influence of postmodernism started to fade as we moved towards the twenty-first century, what can we say about the world view it evolved into?

There have been many attempts to name what came after postmodernism, including, naturally enough, post-postmodernism. Other suggested names have included 'post-digital', 'performa-tism', 'altermodernism', 'post-irony', 'cosmodernism', 'New Sincer-ity', 'digimodernism' and 'post-humanism'. The name that seems to be sticking, however, is 'metamodernism'.

One problem with this name is that it is easy to assume the 'meta-' prefix implies a self-referential aspect. This is not the case. Instead, it comes from the Greek word *metaxy*. Plato defined *metaxy* as being between two contrasting poles. Mankind's *metaxy*, Plato explained, was to be between the animal world and the immaterial world of spirits and gods. The idea is an important one in ancient Greek thought. The political philosopher Eric Voegelin described Greek heroes as oscillating 'between life and death, immortality and mortality, perfection and imperfection, time and timelessness, between order and disorder, truth and untruth, sense and sense-lessness of existence.' The way that post-Millennials can oscillate between being childlike and sensible, or being conservative and liberal, is a more contemporary example of *metaxy*. Another meta-modern attitude is accepting that individual data is flawed and big data is still useful.

For the Manchester-born artist Luke Turner, the author of *The Metamodernist Manifesto*, it is important to oscillate between opposing poles rather than choosing one and firmly identifying with it. He claims that 'metamodernism shall be defined as the mercurial condition between and beyond irony and sincerity, naivety and knowingness, relativism and truth, optimism and doubt, in pursuit of a plurality of disparate and elusive horizons.' In his essay 'Metamodernism: A Brief Introduction', Turner wrote that 'the discourse surrounding metamodernism engages with the resurgence of sincerity, hope, romanticism, affect, and the potential for grand narratives and universal truths, whilst not forfeiting all that we've learnt from postmodernism [...] The metamodern generation understands that we can be both ironic and sincere in the same moment; that one does not necessarily diminish the other.' Metamodernism is nothing if not bittersweet.

Turner is best known for his work with the Finnish artist Nastja Säde Rönkkö and the Hollywood superstar Shia LaBeouf, who in turn is best known for his roles in the *Transformers* franchise and for playing Harrison Ford's son in *Indiana Jones and the Kingdom of the Crystal Skull*. This trio, working under the name LaBeouf, Rönkkö & Turner, are responsible for performance art pieces including LaBeouf's red-carpet appearance at a 2014 Cannes Film Festival premiere with a paper bag on his head, and the words 'I am not famous anymore' written on it. LaBeouf followed this by staging a six-day performance called #IAMSORRY, in which he sat crying in a gallery wearing a tuxedo and the paper bag, while visitors abused him in any way they saw fit.

Performances such as this were mocked mercilessly by my generation, who are not comfortable with sincerity and often fail to recognise it when they see it. As the Generation X comedian Stewart Lee once noted, the ultimate taboo in stand-up comedy is 'a man trying to do something sincerely and well'. LaBeouf, Rönkkö & Turner's work requires the understanding that it is both heartfelt and sincere and also a vacuous aspect of celebrity ego at the same time.

LaBeouf, Rönkkö & Turner's work may have left many baffled, but that doesn't mean metamodernism is an elitist intellectual concept that you need to be educated about to understand. A good example of this is Childish Gambino's deeply metamodern 2018 music video *This Is America*. Gambino, the musical pseudonym of Donald Glover, was relatively successful before this video dropped, yet no one expected the extraordinary level of attention it received. Twenty-four hours after being uploaded to YouTube, it had been watched 20 million times. Countless online blogs and think pieces sprang up to dissect it. For all that it could be endlessly analysed, it immediately resonated with mainstream culture in a way that needed no explanation.

A key theme in *This Is America* was how America loves black culture but rejects black people, and how, by singing and dancing for the entertainment of others, black people remain as complicit in this system as the old 'Jim Crow' stereotype. In pre-metamodern days, you can imagine a song that was a clear condemnation of this situation, and you could also imagine how that song's good intentions could be marred by a tendency to preachiness. But in *This Is America*, Glover both condemns and embraces the situation. On first watch, it is the brilliance of his choreography that grabs your attention. He uses African and Jim Crow-related movement and posture to dazzle and entertain us. It is only on further watches that your eyes begin to leave Glover and take in the wider politics happening around him. In the eyes of the Baby Boomer BBC arts reviewer Will Gompertz, this could be 'quite possibly seen as hypocritical'. But to metamodern eyes, Glover using the strengths of the thing he was condemning made total sense, because in doing so the end result was more powerful and effective.

Academic discussions often stress that they are not trying to promote or encourage metamodernism but are simply trying to put a frame around the modern world view. The cultural theorists Robin van den Akker and Timotheus Vermeulen write that 'we wish to state very clearly that we are not celebrating the waning of the postmodern – nor, indeed are we pushing a metamodern agenda', noting that the concept is 'shot through with productive contradictions, simmering tensions, ideological formations and – to be frank – frightening developments (our incapacity to effectively combat xenophobic populism comes to mind)'. Such is the world that Generation Z are attempting to navigate.

Metamodernism has little interest in the still point of the centre. It is not the best of both worlds. It is an exploration of what aspects of extremes could be useful. This is probably most apparent politically.

In the years after the Second World War the institution of the nation state was at its zenith, both as a source of personal identity and as a means of protecting its citizens. As the British Indian novelist Rana Dasgupta has written, 'For a few decades, state power was monumental – almost divine, indeed – and it created the most secure and equal capitalist societies ever known.' That fell apart during the final quarter of the twentieth century, when deregulated financial systems, globalisation, the growth of multinational corporations and the rise of the internet eroded the power of the nation state. The vast majority of wealth being created by the workers of the world started flowing to a tiny number of transnational billionaires, and there was little that nation states could do to stop this.

Anger at this situation led to a rejection of the centrist political position that had dominated during the last years of the twentieth century, because centrists had sat back, allowed it to happen, and had nothing to offer in terms of fixing the problem. Instead, there was an embracing of left-wing and right-wing positions that had previously been seen as extreme. The right attempted to fix the position by reinforcing sovereignty and the power of nation states, while the left attempted to strengthen workers' positions at the expense of corporations. Both positions have been criticised as attempts to roll the clock back to a time that has now gone, and both can be seen as flawed and potentially dangerous.

Such is the metamodern world: we oscillate between extreme opposites because the seas are turbulent and the still point at the centre is no longer an option.

9.

For all that artistic and academic discussions of metamodernism have their value, I suspect that what has changed is simpler and more fundamental than a fashion for oscillating between opposites. I think that at the heart of the post-Millennial's metamodern culture is a shift in the way we view things that are flawed.

For mid-twentieth-century Baby Boomers, finding a flaw in something was reason to dismiss it. Much political discussion, for example, focused on detailing a flaw in your opponent's position and using this as a reason to reject them. Post-Millennials, in contrast, have grown up with access to sufficient information to understand that everything is flawed from some perspective.

If you went around rejecting everything that was flawed, you'd have nothing left. The important question then becomes which, of the various flawed concepts, can you use or combine in a useful or positive way?

You can see how this change developed if you look at the impact of postmodernism on those who lived through it. Baby Boomers grew up before the arrival of the postmodern world and were taught in childhood to believe that absolute truths both existed and could be known. For them, the huge plurality of viewpoints that followed, where perspectives were understood to be limited and relative, was a horrible development. As they understood the world, it was self-evident that these differing positions must be wrong, not equal. It thus became necessary to argue for their favoured 'true' position at the expense of all others, because if one thing was believed to be valid then logically everything else must be invalid. As no one has been able to convince the rest of the world that their favoured perspective is the one absolute truth that all should share, this has led to a lot of anger, disagreement and fundamentalism.

The Baby Boomers' desire for certainty was less evident in Generation X, who were raised in the postmodern world. If there was a great truth to be fought for, we Gen Xers didn't pretend to know what it was. Everything looked equally flaky to us. You can understand why we retreated into irony and cynicism.

Millennials and Generation Z arrived after postmodernism, so they were not raised to expect absolutes in this world. They were raised in a culture where every position was understood as being limited or flawed in some way. When no position is the great unarguable truth, then the fact that all things are flawed now becomes unimportant. Other people's beliefs are no longer fallacies that need to be fought and defeated. The ideological approach to truth starts to give way to a practical approach. Forget what's ideologically pure – what matters is whether or not something is useful in the here and now.

A nice example of this way of thinking comes from the world of physics. Two of its major models, quantum mechanics and relativity, contradict each other. Physicists, however, don't then dismiss these theories because they must be wrong in some way. They recognise that those models are still both powerful tools when applied in the right circumstances. Satnav systems, for example, rely on both quantum mechanics and relativity to work, even though they are incompatible. Pure-minded philosophers may baulk at the contradiction involved, but then pure-minded philosophers would never be able to get a functioning satellite navigation system up and running.

This practical acceptance of our non-ideal world is now visible in everything from political ideology to cultural appreciation or spiritual beliefs. When I was a teenager, music was extremely tribal and used as a definition of identity. In the 1980s, a goth would not admit to liking dance music, just as mods and rockers kept apart in the 1960s and punks and skinheads were mortal enemies in the 1970s. By the twenty-first century, the idea that one musical genre should be held up above all others was over. We had our favourites, of course, but it was accepted that a single genre was not sufficient to fully express every aspect of our personalities.

To give another example, the Canadian Baby Boomer psychologist turned self-help author Jordan Peterson has frequently argued against the practice of identity politics. His arguments against this are numerous, but a prominent one is the accusation that they are arbitrary. Gender, ethnicity and sexuality are only a fraction of the almost unlimited number of categories that you can divide individuals into. People could equally be divided by class, height, emotional intelligence, wealth, family stability, physical strength, nervousness, reliability, levels of creativity, and so on, and it is reductive to people as individuals to define them by just their skin colour or sex lives. In this, it is hard to deny that Peterson is right. The concept of identity politics is indeed inherently flawed. But that doesn't change the fact that it is extremely useful. Looking at society through the lens of gender, ethnicity and sexuality has revealed deeply ingrained prejudices and inequalities. When you look at the impact of movements such as #MeToo and #TimesUp, and the increasing unwillingness to keep quiet about sexual abuse or casual racism, identity politics appears to be doing a lot of good. That is more important than whether it can be held up as something intellectually pure.

To the Baby Boomer intellectual Jordan Peterson, identity politics should be dismissed because it is flawed. To the metamodern generation, it should be embraced so long as it remains useful and produces positive results, and the fact that it is flawed is not really an issue when everything is flawed in one way or another. This is the attitude of a culture that has come through postmodernism and taken its insights on board, rather than the attitude of those who wanted to reject postmodernism because they did not like its criticisms of absolutes.

A useful way to think about this is to recall something you have fallen back in love with, be that a band, an artist or even a person. When you first discovered a favourite band, you found part of what they do immediately attractive or intriguing. This is what held your attention as you got to understand their body of work better. In time, however, you saw past what appealed to you in the first place and began to see the band's limitations. This changed how you thought about them and stopped you loving them quite so unquestioningly. It could lead you to stop listening to their albums, perhaps even for decades. But, in time, if you then went on to revisit them, you did so aware of both their appeal and their weaknesses. Your view of them was based on a full, rounded understanding of who they were, both good and bad, not just a fixation on their superficial appeal. If you found that you still loved them after that, then it was a deeper love than before. Once you had the level of understanding that comes from knowing both the strengths and weaknesses of something, you would not want to go back to insisting that only the good exists.

This attitude, I think, is what lies at the heart of the cultural changes which the name 'metamodern' tries to capture. The ability to oscillate between irony and sincerity, optimism and doubt, or modernism and postmodernism, which definitions of metamodernism tend to focus on, comes from the understanding that all these options are now on the table. You wouldn't want to rely on any of these approaches all the time, because no option is correct in all circumstances. But there are moments when they are the right tool for the job. Switching between different models also provides insights and ways forward that rigidly staying loyal to a single perspective will never reveal.

Once you start looking out for examples of this change in thinking, you will find it almost everywhere. The way in which younger Christians take a very different view to older members of their congregation regarding female priests is one example. To the younger generation, the church is automatically understood to have its flaws, just as everything else does, so to them the idea of trying to fix its mistakes is no longer controversial. Another example is the willingness of younger environmentalists to accept that nuclear power has an important role to play in the fight against climate change, much to the horror of long-established environmental organisations.

In cases like these, the practical consideration of finding the best way forward overrides accepted positions of intellectual purity. To a fundamentalist, living in a metamodern world must be horrible.

4. THE METAMODERN GENERATION

Many of the convulsions we are seeing in fundamentalist religious, political and cultural movements around the world are reactions to our increasingly metamodern culture.

10.

In his 2018 essay 'Post-authenticity and the ironic truths of meme culture', the researcher and analyst Jay Owens sums up today's youth culture in two words: 'Sensibleness, and Memes'. The contradiction, he points out, is that this description is both 'Seriousness, and taking nothing seriously.' Once you have started to recognise metamodernism, this seeming contradiction is not so surprising.

Owens goes on to analyse why meme culture is so big with Generation Z. He concludes that it comes from a place of stress and anxiety. The flippancy and irony of memes make them a safe way to express difficult feelings. As Owens writes, 'through humour, and exaggeration, and irony, a kind of truth emerges about how people are feeling. A truth that they may not have felt able to express straight [...] The formal properties of the meme make it a particularly effective format for delivering an indirect payload of empathy.' Sharing a meme around their friendship group is a way of expressing how teenagers are feeling, and by liking and sharing that meme their friends are signalling that they understand and that they feel that way too. The use of irony like this is a step beyond Millennial culture's love of authenticity. That Generation Z have found a valid, practical use for irony shows how much they have progressed over Generation X.

If this combination of empathy, anxiety and the metamodern use of flawed extremes to achieve goals can be said to define current teenagers, then the speech by a teenager, Samantha Fuentes, at the Washington March for Our Lives Rally tells you everything you need to know about this generation. Fuentes was a student at Marjory Stoneman Douglas High School in Parkland, Florida. She was shot in both legs when that school was attacked by a mass shooter in February 2018. She now has shrapnel embedded behind her cheek and eye which, because of its location, will probably never be removed.

Fuentes was giving a passionate, televised speech in front of an enormous crowd at the gun-control rally when nerves caused her to vomit on stage. After recovering, she proclaimed, 'I just threw up on international television and it feels great!' and then launched straight back into her speech.

The nerves and anxiety that caused her to be physically sick are the reason these kids have been labelled 'generation snowflake', but clearly that is not the whole story here. Anxiety is constant worry, and this comes as part of a package with an increased level of empathy. That empathy demands action after seeing 17 school-mates murdered, because the thought of others going through what you have experienced is too painful for an empathetic person to accept.

But what response was possible? Adult American culture looked set to ignore the problem, as it had done after all the previous school shootings. Politicians were clearly not going to do anything. Fuentes received a call from President Trump when she was in hospital. 'Talking to the president, I've never been so unimpressed by a person in my life. He didn't make me feel better in the slightest,' she has said. She recalled how the president 'said "Oh boy, oh boy, oh boy", like seven times'.

Coming together with your teenage peers and organising one of the largest mass protests in American history might, on the face of it, be seen as an extreme response. In some ways it was far from ideal, as it inevitably made the issue a tribal, politicised one, and this guaranteed that these traumatised kids would receive an extraordinary level of abuse. Yet what other tools did these children have that would produce results? How else could they overcome social inertia and mobilise the required support of the electorate? The centrist approach of reasoned discussion and compromise was clearly not going to achieve anything, as the history of American school shootings affecting the Millennial generation has shown. To the metamodern Generation Z, the hard work, clarity of purpose and dedicated activism displayed by the Parkland survivors was the only way forward that might achieve anything.

The 'generation snowflake' anxiety that caused Sam Fuentes to puke so publicly, therefore, has to be seen as part of a larger picture. It makes their achievements almost more remarkable, because the only thing stronger than their anxiety was their empathy for future victims. Bravery is only possible in those that are afraid. This generation are scared, they are nervous, and they would far prefer not to have the battles in front of them that they will have to face, but by God they care about what happens to other people.

I am reminded of how much this generation suffer from anxiety as I go into the local Co-op to get a loaf of bread. The Co-op is near a large sixth-form college, so it is often full of teenagers. They move round the shop in packs, gossiping loudly, as if unaware they are in a public place. The shrieks and hubbub they emit seem an octave higher than adult conversations. They queue for the automated tills, while I wait in line for the manned ones.

An alarm sounds. There is a problem with the automatic tills. A member of staff goes over to deal with the issue while I pay for my loaf. It becomes apparent that all the automatic tills have become unable to take card payments, so a shop worker asks the teenagers to pay at the manned tills until the problem is fixed. The teenagers don't seem happy about this but, seeing no alternative, they walk over to the manned tills. All except one, that is. One girl, aged about 16, looks a little panicked by this situation. Her eyes dart around the shop, unsure what to do. Then she turns and heads for the exit, dropping the item she was planning to buy in the newspaper stand next to the door as she does so.

I collect my change and leave. Out of curiosity, I look at the newspapers to see what item the teenager had been planning to buy. It was a bag of pink and white marshmallows. The paper it sits on, I notice, has an account of the Washington gun-control march organised by teenagers who survived the Parkland school shooting in Florida. Up to 2 million people participated across the United States, making it one of the largest protests in American history. This was the protest at which Samantha Fuentes puked on stage through nerves but delivered her speech regardless. These teenagers were the generation that I had come to admire and put my faith in. These were the empathetic, networked, metamodern kids who will adapt to the changing climate, see the world's problems clearly, and tackle them with more wisdom and humanity than those of us still contaminated by twentieth-century thinking. And there, on top of an account of their history-making achievement, was a packet of marshmallows that they wanted, but were too socially anxious to buy.

We are not in a tragedy, I realised at that moment. I'm sure I can hear a laugh track. This is definitely a comedy.

5.
THE DREAM OF SPACE

1.

An old Zen kōan asks, if a tree falls in the forest and no one is around to hear it, does it still make a noise? For modern scientists and Zen masters alike, the answer to this is no, it does not. The impact would cause ripples to radiate through the air from the point of impact outwards, but these ripples are not a sound. A sound only occurs in the mind, after those atmospheric waves have been detected by the ear, transformed into signals, passed through our nervous system and registered by our consciousness. On 6 February 2018, a more modern version of that thought experiment appeared: if a car stereo plays David Bowie's 'Space Oddity' on an endless loop while in orbit around Mars and no one is around to hear it, does it make a sound?

The car in question was launched by the South African billionaire businessman and engineer Elon Musk. As well as creating SpaceX, the company that launched the car into space, Musk is also CEO of the electric car company Tesla. The $100,000 Tesla Roadster currently hurtling through space was his own, the first

Tesla off the production line. On a circuit board inside the car is written, 'Made on Earth by humans'.

At the press conference immediately after the launch, Musk appeared stunned, as if he was still trying to process exactly what it was that he had just done. The launch had been the maiden test flight of the Falcon Heavy rocket. This can carry three times more weight than his previous Falcon 9 rockets and is intended to bring Musk's lifelong dream of colonising Mars one step closer. A test flight needed cargo as ballast, and Musk had had the idea of using his own car. 'It's kind of silly and fun, but I think silly fun things are important,' he explained to journalists. 'Normally for a new rocket they launch a block of concrete or something like that, I mean that's so boring. I think the imagery of it [the car in space] is something that's going to get people excited around the world. And it's still tripping me out. I'm tripping balls here.'

The image of the car in space was incredibly important, as the visibly shaken Musk understood. This was evident from the comments of those watching the live stream of the car orbiting the Earth. While the expected space enthusiasts were present, asking informed questions about the fate of the rocket's central thruster, they were massively outnumbered by non-space geeks who had been passed the link by online friends. These were people who never usually watch live streams of rocket launches and were trying to understand exactly what they were seeing. 'This dude's sent his car into space!' was a typical reaction. Others were in denial ('Fake!') or just simply amazed ('This is the coolest fucking thing I've ever seen!'). Many thought it didn't look real, as if it were cheap CGI. This was understandable, because we are not used to seeing how light behaves with no atmosphere to scatter it. Musk himself also commented on this effect. 'You can tell it's real because it looks so fake,' he said. 'Honestly, we'd have way better CGI if it was fake.'

Some online commentators reacted with customary knee-jerk cynicism, but the sneering was noticeably half-hearted. The writer and activist Naomi Klein made perhaps the best stab at it by tweeting 'About Elon's big day . . . this is a car commercial in space. Everyone: pls stop participating.' But after seeing the reaction to her comment, she soon followed it up with 'Everyone hates my tweet it seems. I get it, we need moments of awe. But between his union busting at the Tesla factory and the incredibly dangerous flamethrowers, I think this guy is out of control. He's a savvier Branson and that ended badly.' The union practices at Musk's companies are a legitimate issue, but they and his interest in flamethrowers paled in comparison to the significance of what he had just done. He was rewriting the circumambient mythos in the minds of millions. He was suggesting that we did have a future after all.

The stroke of genius was the mannequin, dressed in a spacesuit, which sat in the driver's seat. If it had been empty, the car would have looked like it had been discarded by an owner with more money than sense. But with the fake driver in place, the visual story was completely different. This was a journey. With the top down in a sporty roadster, someone was driving a car to Mars. Institutions like NASA had tried for decades to get people excited about the idea of mankind travelling to Mars, but the public had never really bought into it. The activities of NASA seemed too far removed from their own lives for them to be able to relate to the idea. Yet the idea of sitting in a nice sports car with David Bowie on the stereo while driving to Mars was considerably more relatable. If the evidence of their own eyes was anything to go by, it was now entirely possible. Suddenly, in the minds of millions, the idea that we would go to Mars became real.

The image sparked all sorts of thoughts and wild notions. The comics writer Warren Ellis joked, 'I'm pretty sure the Starman in the car in space is a mannequin but what if Jeff Bezos [Musk's billionaire rival] suddenly mysteriously disappears and nobody knows why?' This remark was the product of a great storyteller's mind. Could anyone dream up a better way to dispose of a body in plain sight, in such a way that proof could never be found? A conspiracy theory also started circulating that the mannequin was actually the body of David Bowie, whose real funeral arrangements had been kept private, and that with this final act the album *Blackstar* now made sense.

Will we ever see that car and mannequin again? SpaceX's scientists believe it could remain in Mars's orbit for up to a billion years. That's a long time, and humanity could get up to a lot in those years. If you recall the excitement when the sunken Tudor warship the *Mary Rose* was recovered after 500 years at the bottom of the English Channel, or the thrill that came when the wreck of the *Titanic* was found and filmed, you can start to imagine the aura that Elon Musk's car will have in a thousand years or so. Will we be regularly travelling between planets at that point? Will a future treasure hunter reclaim the vehicle and settle for all time the question of what happened to the remains of David Bowie? The visual image of the relaxed mannequin driving his roadster to Mars makes us think that this is a journey we will take many times in the years to come, but how true is this?

If a car stereo plays David Bowie's 'Space Oddity' on an endless loop while in Mars orbit and no one is around to hear it, does it make a sound? The answer to this is most definitely not. Not only is there no one there to hear the ripples and experience them as sound, but there are no ripples. The car is in the vacuum of space. Although its stereo's speakers are working as intended, no sound waves can be emitted out of them in the absence of an atmosphere. The only way 'Space Oddity' is playing is in our imaginations. In our minds, that sound system is crystal-clear, and the song has never sounded better. That mannequin is hearing the song in a way that we can only dream of. The song is a beautiful fantasy. It is an ideal metaphor, then, for our future in space.

2.

I don't recall watching *Star Trek* as a kid, but it didn't really matter whether you watched it or not. *Star Trek* seeped into wider culture, an ever-present future myth that reached you regardless of whether you tuned in to the show. Everyone knew who Captain Kirk and Mr Spock were, with their brightly coloured jumpers and too-short trousers. Everybody could recognise their strangely elegant starship, the USS *Enterprise*. This pop culture imagery came with a promise: we did have a future, it assured us, and it would be an exciting one.

In the twenty-third century that *Star Trek* promised, money, conflict and bigotry were all products of the past. Mankind had matured to the point where we were able to leave those damaging, destructive concepts behind. Our reward was the ability to explore the galaxy in peace, encountering alien cultures and discovering strange new worlds.

It all seemed plausible. As Martin Luther King promised, quoting the American clergyman Theodore Parker, 'the arc of the moral universe is long, but it bends towards justice'. We were becoming better people. The human story was a story of exploration, from the settling of the American frontier or the centuries of colonial exploration, all the way back to our first journeys out of Africa at the dawn of prehistory. Technology was advancing at an accelerating rate, and it would presumably continue to do so. In the 1960s, the notion that we would conquer our problems and travel the stars exploring the final frontier together seemed like a safe bet.

Visually, the starship *Enterprise* was a product of mid-twentieth-century America. It had a round disc section at one end, like a B-movie UFO, balanced by a pair of long thin engines that swooped upwards at the back like the tail fins on a 1950s Cadillac. It was created at the birth of colour television, when cinematographers were going hog-wild with their use of colour. Every corridor or blank wall was flooded with purple or blue light. This would have distracted from the actors in the foreground, had they not all been dressed in bright primary colours. Visually, *Star Trek* was a joy.

The programme's central hero was Captain James T. Kirk, an all-American man of action. Kirk was entirely confident and decisive, and untroubled by doubt. He always knew exactly what to do and his crew trusted him completely. With men like Kirk around, exploring the galaxy was not just something that mankind might attempt, but something it would be great at.

Star Trek was born during a decade of race riots, the escalating war in Vietnam and Russian superiority in space. The Soviets had chalked up an extensive list of successes beyond the Earth's atmosphere. They had launched Sputnik, the first artificial satellite, and they sent the first animal, man and woman into space. They were confidently performing space walks and sending probes around the moon while the Americans were watching rocket after rocket explode on the launch pad. *Star Trek* was a brightly coloured promise made at a time when the geopolitical outcome was far from certain, but a promise which reality then began to live up to. The series was so confident that America was destined to conquer space that you might never guess its final episode was broadcast before *Apollo 11* first put a man on the moon.

The series ran from 8 September 1966 to 3 June 1969. It was not a success during its initial run, rating poorly and being in frequent danger of cancellation. It was, perhaps, ahead of its time. At its core was a message of ethnic inclusivity, with a crew containing officers of African, Japanese, Russian, Scottish and alien descent. For the programme's creator, Gene Roddenberry, this vision of inclusivity was the whole point of the show's existence. Judging people on their skin colour or geographic background was so self-evidently foolish that presenting a future where we hadn't outgrown such behaviour struck Roddenberry as ludicrous. When he received a letter from a TV station in the American south saying that they would not show the programme because it included a black officer, his response was 'Fuck off, then.'

As Roddenberry explained, 'I had been a freelance writer for about a dozen years and was chafing increasingly at the commercial censorship on television, which was very strong in those days. You really couldn't talk about anything you cared to talk about, and I decided I was going to leave TV unless I could find some way to write about what I wanted [...] It seemed to me that perhaps if I wanted to talk about sex, religion, politics, make some comments about Vietnam, and so on, that if I had similar situations involving these subjects happening on other planets to little green people, indeed it might get by, and it did. It apparently went right over the censors' heads, but all the 14-year-olds in our audience knew exactly what we were talking about.' Making TV about the twenty-third century, Roddenberry realised, was the only way he could make television about the twentieth.

The show's philosophy, as first laid down by Roddenberry and developed by producer Gene L. Coon, is summarised in the show itself by the acronym IDIC, which stands for Infinite Diversity in Infinite Combinations. According to IDIC, variety is a virtue and the source of truth and beauty. Under this philosophy all life is equally precious, and there is a place in the universe for everyone. Roddenberry understood that the future is not just a story of changing technology. It is a story about how humanity will change. The spaceships, transporters and phasers are a fun backdrop for all this, but they were not what was important.

To modern eyes, the original *Star Trek* can seem conflicted. It tells of a time of peaceful exploration after humanity has evolved beyond war, yet it focuses on a military crew punching and shooting their way around the galaxy. It talks of equality, but its portrayal of women is considerably less progressive than its treatment of ethnicity. Fred Freiberger, the producer of the third series, summarised it as 'tits in space'. Despite all his progressive ideals, Roddenberry was, according to Gene L. Coon's assistant Ande Richardson, 'a sexist and manipulative person who disregarded women'. The original series was neatly summarised by the American science fiction writer Ed Naha as 'an intellectual and emotional refuge for people who believed in positive change. And cleavage.'

All this needs to be seen in the context of the times. In the original pilot, Roddenberry cast a female officer as the Captain's second-in-command, but this idea was scrapped after test reports showed that female viewers hated it. A typical comment about the character was, 'Who does she think she is?' Yet even given the context of the times, the programme does have moments that now seem particularly inexcusable. In the final episode, for example, after Kirk and a female scientist swap bodies, we are told that 'it is better to be dead than to live alone in the body of a woman'.

For all its faults, *Star Trek*'s tales of friendship and adventure proved to be an ideal delivery mechanism to inject its philosophy of equality into the wider culture. Its audience may have initially been small, but those who did watch heard what it was trying to say. The message resonated deeply. Slowly and perhaps unconsciously, its vision of the future was absorbed into the wider culture. The idea that mankind was destined to leave the planet and travel among the stars, where no man had gone before, went from being a dream to an assumption. It provided the necessary optimistic counterweight to the idea that mankind would wipe itself out in a nuclear holocaust.

It helped that our previous visions of the future were becoming outdated and increasingly implausible. For most of our history these had been provided by religion, but tales of Ragnarok, Apocalypse or Second Comings were unconvincing in the technological age. The early twentieth century generated a number of replacement visions, from H. G. Wells's Socialist World State to Adolf Hitler's Thousand-Year Reich, but none of these held a great deal of appeal after the Second World War. Almost by default, as Neil Armstrong set foot on the moon and global co-operation created the International Space Station, Roddenberry's dream became the imagined destination of the West.

At the time of writing, the latest television incarnation of *Star Trek* is being broadcast by Netflix. *Star Trek: Discovery* is set about ten years before the voyages of Kirk, Spock and McCoy. It contains none of the bickering camaraderie that marked those three as friends. Humanity is at war and the main character is a mutineer who is given a job experimenting on animals on a secret military weapons research ship. The captain of the *Discovery* has a mysterious problem with his eyes that causes him to skulk around moodily in dark, shadowy rooms; there is no vivid blue and purple lighting in this ship. Instead, we have torture, deceit, mistrust, post-traumatic stress disorder, which was initially implied to be caused by male rape, and use of the word 'fuck'. It is a series that feels contemporary. It tonally matches other modern sci-fi series like *The Expanse* or *Altered Carbon* by presenting a bleak, miserable future that few people would want to live in. Although the camaraderie between characters was increased in the second season, only traces of Roddenberry's optimistic vision still remain. In terms of future *Star Trek* films, meanwhile, the studio has begun developing scripts with Quentin Tarantino. To go from Gene Roddenberry to Quentin Tarantino is quite a significant change.

What has changed in the past 50 years? Star Trek has gone from being an inspiring vision of a future Utopia that you longed to live in, to a depressing future you hope you'll never see. If even *Star Trek* can no longer take the founding ideals of *Star Trek* seriously, then it is not surprising that the rest of culture has given up on them. So why has *Star Trek* rejected itself?

3.

Roddenberry lost control of *Star Trek* on three separate occasions. The first was back in 1968, when the third and final season of the original series was going into production. Roddenberry threatened to walk away from the series if it was screened at the later time of 10 p.m. on Fridays, and the network decided that he should do just that.

The second time Roddenberry lost control of *Trek* was after the release of *Star Trek: The Motion Picture* in 1979. This was a big budget, effects-heavy affair, which the studio green-lit following the success of George Lucas's *Star Wars*. The story itself mirrored Roddenberry's return to *Trek*, because it was about an immense, bewildering and largely misunderstood space-bound phenomenon returning to Earth to reunite with its creator.

Roddenberry had spent the 1970s encouraging and nurturing the emerging *Trek* fan culture, which was then a surprising and unexpected phenomenon. Nowadays it is unremarkable for an adult to say they are a fan of genre shows such as *Game of Thrones*, *The Walking Dead* or *Doctor Who*, but being an adult fan of *Star Trek* in the 1970s was a cause of serious social embarrassment. TV shows set in unreal worlds were understood to be children's entertainment. That childhood was lengthening and the boundary with adulthood was blurring was not understood in the 1960s.

Perhaps as a reaction to this, *Trek* fans talked up the aspects of the show that were intelligent, philosophical and optimistic. They downplayed the fun, cheesy, adventure serial aspects of the series, even if those were the things that caused them to love it in the first place. As Roddenberry toured the emerging fan convention circuit, trying to understand why his show had spawned this unprecedented social movement, he kept hearing fans talk about the importance of the idealistic, positive *Trek* vision. This, he decided, was why the programme worked. Even though the original series had at times been a very different beast to how it was becoming memorialised, Roddenberry bought into this worthy caricature of his show. It was this that he realised onscreen in *Star Trek: The Motion Picture*.

The movie's development did not go smoothly. The film entered production with the script still in flux. Problems with the special effects caused the budget to balloon from $15m to a then unheard-of $46m. The resulting film was still a success at the box office, but it was critically accused of being too long, too ponderous and too humourless. It lacked the action and rich character interplay that made the original series so enjoyable, and it attracted unflattering nicknames, such as *Star Trek: The Slow Motion Picture* or *Star Trek: The Motionless Picture*. As the studio saw it, the problem was Roddenberry. He had spent $46m on making a film for stoners and die-hard fans instead of something that would please the millions of casual viewers now tuning in to *Star Trek* reruns on TV. The film's box-office success showed there was a demand for more *Trek*, but the studio was not going to trust Roddenberry to supply it.

Star Trek II: The Wrath of Khan was released in 1982. The story was *Moby Dick* in space, with Captain Kirk playing the role of the white whale and Ricardo Montalbán, reprising his role as the villain Khan from the original series, playing a crazed galactic Ahab. It cost less than a quarter of the first film to make and it played up the naval and military aspects of the original series, which Roddenberry had wanted to play down. As Nicholas Meyer, the director and uncredited writer of *Star Trek II: The Wrath of Khan*, explained, 'Roddenberry definitely averred or opined that *Star Trek* was not a naval operation, not a military operation, it was a sort of Coast Guard, is how he put it. And I thought watching these episodes that that didn't seem to be the case. This was definitely a form of gunboat diplomacy.'

Meyer's vision of the future was fundamentally different to that of Roddenberry. 'It doesn't seem, aside from certain technological advancements, the world has substantially changed,' he has said. 'People are still making fists and cutting off each other's heads. And, in that sense, my version of *Star Trek* was a gloomier, darker version. But the people were still the same people. They were just having to confront a less optimistic reality.' Meyer simply didn't believe in the core idea at the heart of Roddenberry's *Star Trek*, which was that people will change for the better.

This difference of opinion is captured neatly by an argument between Meyer and Roddenberry regarding a 'No Smoking' sign on the bridge of the *Enterprise*. As the producer Ralph Winter recalls, '[Meyer] and Roddenberry went round and round about no smoking because, Roddenberry said, "No one is going to smoke in the twenty-third century." And Nick said, "People have been smoking for hundreds of years and they will for hundreds of years." And they were at it on the set. They were arguing and yelling, it was bizarre. But that was Gene's view of the future and utopia and what he believed.' In this instance, history seems to be proving Roddenberry right and Meyer wrong. In 1965, 42 per cent of Americans smoked, but by 2016 that number had dropped to

15.5 per cent. The idea that no twenty-third-century astronaut will smoke cigarettes seems entirely plausible.

Ultimately, Meyer thought that a 1980s audience didn't want a vision of a future Utopia. They wanted solid, escapist entertainment that echoed the world as it was then. And, in this, he has largely proven to be correct. The critics and fans loved the film. It went on to make a lot of money and it is still widely regarded as the best *Trek* film. Roddenberry hated it.

The third time that Roddenberry lost control of *Trek* was during the sequel TV series *Star Trek: The Next Generation*. The ongoing success of the film series showed that the audience for *Trek* was not going away, but the ballooning salaries and advancing years of stars William Shatner and Leonard Nimoy made a new approach desirable. Roddenberry had the answer. He would move forward from the twenty-third century to the twenty-fourth. A new TV series would follow the adventures of an entirely new crew on a larger and more advanced starship *Enterprise*. This allowed him to dream up more advanced technology such as the Holodeck, in which cast members could enter virtual simulations of whatever world they wished. Roddenberry made a point of downplaying the naval or military nature of this new vessel. There were families among the crew, and their living conditions were luxurious. It may have been a spaceship run under a naval hierarchy, but it now had carpets.

By the twenty-fourth century, Roddenberry insisted, mankind would have overcome its demons to the extent that there would no longer be any arguments between the members of his crew. To hammer the point home he added a counsellor as a major crew member, who was given a chair on the bridge next to the captain himself. Her job was to head off any disagreements before they got started. This caused a great deal of consternation among the writing staff. If the cast have to remain best buddies throughout, where would the drama come from?

Buried inside this problem is a deeper concern. A boring TV drama is one thing, but what are the implications for the actual future that's approaching? If we do solve our problems and build a fair, just and equal Utopia, will we find life boring?

4.

Literature is littered with dramatic dystopias, but stories about Utopias are fewer and farther between. Those that do exist are rarely celebrated. When we think of an author like Aldous Huxley, we are far more likely to talk about his dystopian novel *Brave New World* than we are his utopian novel *Island*.

Perhaps the most successful attempt at a science fiction Utopia is the *Culture* series of novels by Iain M. Banks. They depict a future where the interstellar economy is run automatically by machines and vast AI beings called Minds. It is not, strictly speaking, a vision of humanity's future. Earth is visited in one of Banks's short stories and the decision is made not to help it, but to leave it to develop in its own way like a control element in a scientific study. Despite this, Banks's future still works as a potential blueprint for where we are headed.

You would be hard pushed to call Banks's future dull, for it is rich in invention and humour. The names of spacecraft are a good example of this. They are chosen by the ships themselves and are often slightly inappropriate, which does makes them sound as if they really were chosen by AI. They frequently sound more like racehorse names than those of spaceship. Banks's stories contain spacecraft including *You Would If You Really Loved Me*, *What Are the Civilian Applications?*, *All the Same, I Saw It First* and *A Fine*

Disregard for Awkward Facts. There are ships called *Helpless In the Face of Your Beauty, Synchronise Your Dogmas, That's Funny, It Worked Last Time . . .* and *God Told Me to Do It.* Elon Musk has named two autonomous drone ships after Culture spacecraft. These are ocean-going barge ships on which SpaceX rockets land after a successful launch. They are called *Of Course I Still Love You* and *Just Read the Instructions.*

Life in the Culture is idyllic. People spend a great deal of time attending to sensual luxuries, parties and the pursuit of extreme pleasure. They have been biologically altered to contain drug glands, which allows them to modify their mood at will. Hovering drones attend to their every need, like flying butlers. Ageing is a choice, as is work. Banks, when he wants to, can make life in the Culture sound very appealing. Yet his stories almost always involve the utopian Culture encountering or interacting with an outside, non-utopian culture, and the narrative follows the culture clash that ensues. The demands of drama dictate that a fully utopian novel is unlikely to hold our attention. As we have known since the ancient Greeks, it is conflict and resolution that interest us. It was this that led to Roddenberry losing control of *Star Trek* for the third, and final, time.

Roddenberry's insistence that there be no conflict among his crew became a cause of unhappiness among the programme's writing staff, of which there was a famously high turnover. As the writer Hans Beimler said, 'Gene kept thinking that [people in the twenty-fourth century] were better people; his view of the world was, this was a better world and these were better people than we are. And that was a fundamental problem [. . .] It also belied the principle that we [the rest of the writing team] firmly believed, which is people are not going to get better.'

It didn't help that the rules Roddenberry insisted his staff follow did not always seem to apply to the episodes he wrote himself, or that there was an increasing gulf between his stated ideals and the reality of how he lived his life. Numerous staff claim Roddenberry used strong racist and sexist language during *Star Trek: The Next Generation* meetings. The creative consultant, David Gerrold, recalled him referring to black people as 'spearchuckers'. Roddenberry once gave a lengthy and inspiring speech on the importance of female representation, then, as one of the producers, Herb Wright, recounted, 'all of a sudden something kicks in and he changes: "However, we also don't want to infer that it would be a better society if women ruled, because as we all know," and he's getting louder and louder, "women are goddammed cunts! You can't trust them! They're vicious creatures who will cut your throat when you're not looking!" Then he looks out of the window, looks at the outline, and says, "Okay, on page eight..." and continues like that didn't even happen.'

Roddenberry was getting old and becoming increasingly infirm. His previous experience of being removed from *Star Trek* had left him obsessed with control, and this made him seem increasingly paranoid. His use of cocaine and alcohol for much of his life did not help. He was never officially sacked, but his increasingly strange behaviour caused him to be gradually removed from day-to-day production decisions.

With his control slipping away, Roddenberry was unable to prevent an episode of *The Next Generation* that radically altered the tone of the *Trek* universe. In this story, a godlike omnipotent being called Q flung the *Enterprise* to the far side of the galaxy, to demonstrate mankind's hubris. Here they met a soulless, networked species with a hive mind, called the Borg. The Borg were unstoppable, and far too powerful for mankind to ever defeat. Before this episode, exploring the galaxy had been portrayed as something that humanity was good at. Kirk was entirely confident and capable of overcoming whatever space had to throw at him.

In this episode, the crew were reduced to begging the omnipotent Q to save them and bring them home. From that point onwards, mankind could no longer view itself as master of all it encountered. Humanity was out of its depth. Somewhere out there among the stars lurked something that humbled us.

By series four, Roddenberry's status had diminished to the extent that the show could produce an episode which dealt with the troubled relationship between Captain Jean-Luc Picard and his brother. Roddenberry hated the script, for he did not believe that the twenty-fourth-century Picard would have dysfunctional family relationships. The episode showed Picard solving his problems by punching his brother in the face, engaging in mud wrestling, then getting drunk.

On 24 October 1991, during the fifth season of *Star Trek: The Next Generation*, Roddenberry died. He was seventy years old. The Great Bird, as the fans nicknamed him, had left his successful and profitable galaxy, leaving no one in control who believed that humanity would become better. Only a few months after he died, plans were announced for a new spin-off series called *Star Trek: Deep Space Nine*. This was a darker, colder vision of the future, set on a remote outpost during a time of intergalactic war, full of previously forbidden personal conflicts.

The *Trek* universe did nothing to contradict the established continuity of a future Utopia, and the programme makers spoke repeatedly about honouring Roddenberry's ideals. But it would be fair to say that they were honouring the letter of the law, rather than the spirit. Once Roddenberry had gone, the tone of his fictional universe changed.

Ten years after his death, and many hundreds of episodes later, exhaustion set in for the franchise. The twenty-fourth century had been milked for all it was worth. A change was needed, similar to how Roddenberry had moved the story forward by a century in *The*

Next Generation, thereby providing a new moral and technological backdrop for stories. The logical thing to have done would have been to dream up the twenty-fifth century. But there was nobody working on the franchise who had a vision of what the twenty-fifth century would be like.

Unable to imagine going forward, they went back in time. In the twenty-first century the franchise has given us *Star Trek: Enterprise* (2001–05), *Star Trek: Discovery* (2017 onwards) and the rebooted big-budget film franchise. All of these are all set either in the time of the original series or earlier. The *Star Trek* franchise is no longer going where no *Trek* had gone before.

Roddenberry was a flawed, difficult man, but he believed that the vision and promise at the heart of his creation mattered. It was that which elevated it above corporate entertainment and turned it, if only briefly, into a future myth for the West. It keeps going, of course, and it is frequently slick, well produced and entertaining. But the franchise itself no longer believes that we are becoming better people and that our society is improving. It has become a zombie franchise, staggering on even though its heart has long since stopped.

5.

One of the main difficulties involved in coming late to *Star Trek*, and trying to understand its evolution over the years, is that eventually you arrive at the song 'Faith of the Heart'. This is the theme music to the prequel series *Star Trek: Enterprise* (2001–05). The song is a soft-rock ballad, written by Diane Warren and was originally performed by Rod Stewart for the soundtrack to the film *Patch Adams*. The *Enterprise* version is by the English tenor Russell Watson.

5. THE DREAM OF SPACE

I discovered that 'Faith of the Heart' made the act of watching this series difficult, because whenever I put an episode on, a member of my family would choose the moment it played to enter the room. My partner, Joanne, for example, would appear and proclaim, 'Jesus, what is this godawful shite?' Our children don't complain so loudly, but they still give me looks of extreme disappointment.

It is not only my family who have a problem with this song. When the series first aired, fans protested outside Paramount Studios and collected petitions demanding that the song be replaced. One petition read, 'We wish to express our unmitigated disgust with the theme song that has been selected for the new *Enterprise* series, it is not fit to be scraped off the bottom of a Klingon's boot.' The actor and writer Simon Pegg, who plays Scotty in the most recent films, said, 'I think that the theme music to *Enterprise* was probably the most hideous *Star Trek* moment in history [...] I've never seen *Enterprise*, because I couldn't get past that music. It would still be ringing in my ears when the show starts.'

This is a shame, because within those opening titles is a succinct encapsulation of the current post-Roddenberry *Star Trek* future myth. They contain a potted history of humanity's efforts to journey away from the land we live on, presented in rough chronological order. It begins with images of oceans and maps and a Polynesian sailing boat at sunset. These fade into a diagram of the eighteenth-century Royal Navy sloop HMS *Enterprise*, a reminder that there has been a long history of ships named *Enterprise*. Roddenberry had borrowed the name for his starship from the USS *Enterprise* which turned the tide of the Battle of Midway.

The title sequence continues with shots of aviation and space-flight pioneers. We see Amelia Earhart, Chuck Yeager and Alan Shepherd, as well as a succession of vehicles ranging from hot-air balloons and underwater submersibles to Charles Lindbergh's

Spirit of St. Louis monoplane and the experimental supersonic plane the Bell X-1. These heroic individuals, who risked their own lives to go further and faster than anyone before, are presented as part of a tradition stretching back centuries. These are the people, the titles suggest, who pushed back the boundaries which constrained us and allowed the potential of humanity to flower. Then we see footage from the 1970s of the first space shuttle and clearly see its name: *Enterprise*.

This was the moment where history and myth crossed over. NASA's first prototype space shuttle was intended to be called *Constitution* until fans of the original *Star Trek* series organised a nationwide letter-writing campaign. President Ford received hundreds of thousands of letters asking for the ship to be named *Enterprise*, which, he decided, was an appropriate name for an American ship. Many NASA employees and aerospace engineers openly admitted that it was the TV series that inspired their careers, so Ford's willingness to change the name was well received. This started the practice of naming space-exploration vehicles after fictional spacecraft that was continued by Elon Musk's drone ships *Just Read the Instructions* and *Of Course I Still Love* you 40 years later.

The myth then absorbed the reality it had inspired. The 1979 film *Star Trek: The Motion Picture*, released three years after the space shuttle was named, included an image of the space shuttle *Enterprise* as part of a display of the starship's predecessors. The fiction of *Star Trek* took the tradition of sailing vessels named *Enterprise*, projected this forward into the space age and inspired reality to fall in line and make that prediction a reality – such is the power of storytelling. Yet the space shuttle *Enterprise* was an early prototype. It never did leave Earth's atmosphere or enter space. The voyages of the spaceship *Enterprise* are dreams, not reality.

The programme's titles then continue through the early years of space exploration. They include shots of a Saturn V rocket taking off, Buzz Aldrin's footprint on the moon and the *Mars Pathfinder* rover exploring the red planet. This potted history ends with a shot of the International Space Station being constructed. From there on in, the final three shots switch from reality to fiction. They depict space vessels to come, and they end with the first spacecraft capable of interstellar travel, the *NX-01 Enterprise*. The message of the title sequence is clear. Travelling to the stars is inevitable. We've been working towards it for hundreds of years and we're nearly there. Only the last few steps remain.

What is needed to take those last few steps, the song tells us, is faith. It's been a long journey, the lyric admits, but we are nearly there. Nothing can stop us; our dreams will come true and we will reach any star. Provided, of course, that we believe. The magical thinking of faith is all we need to achieve the goal of spreading out across the galaxy. It's an incredibly seductive notion, but it is a classic example of an arrow-flight projection.

Those last few steps are not inevitable or trivial at all. It is not the case that we have come far and have only the last few hurdles to go. The title sequence is like someone who lives on the East Coast of America opening their front door and walking down their path to their front gate, then claiming that they've only got a few steps to go to reach California. The difficulties involved in travelling from the Earth to the stars far exceed the difficulties faced by the pioneering aviators shown in the title sequence. They will take a lot more than faith to overcome.

6.

The main problem with travelling to the stars, as Douglas Adams so neatly put it, is that 'Space is big. You just won't believe how vastly, hugely, mind-bogglingly big it is.'

In 2013 we discovered that the star Kepler-62 in the constellation Lyra was orbited by five planets, one of which was rocky and just the right distance from its sun for life to be possible. This planet, Kepler-62f, is one of the best candidates we have found for hosting alien life. This is very exciting, but unfortunately it is a very long way away. It is so distant that it takes light 1,200 years to travel from here to there. Travelling faster than light is impossible, so if we wanted to fly to Kepler-62f, even if we were able to travel at the fastest speed physically possible, then it would take us 1,200 years to get there and 1,200 years to get back.

The realities of the size of space are a problem for a TV format like *Star Trek*, which requires its crew to arrive at an interesting new planet every week. *Star Trek* fixed this problem by claiming that, in the future, we'll invent a faster-than-light engine called a warp drive. This was pure fictional hand-waving when Rodden-berry dreamt it up but, rather brilliantly, the Mexican theoretical physicist Miguel Alcubierre discovered in 1994 that such an inven-tion was theoretically plausible. It does not necessarily contradict the laws of physics.

His idea, which is known as the Alcubierre warp drive, works by squeezing space-time in front of a spaceship and expanding space-time behind it. This causes the bubble of space-time around your ship to zoom off at incredible speeds across the cosmos. Importantly, the ship itself is technically still. It is the bubble of space-time that it sits in which is whizzing across the galaxy. This means that the spaceship does not contradict Einstein's laws about an object travelling faster than light.

Gene Roddenberry's use of the word 'warp' has inspired the science which made it theoretically possible for mankind to travel to the stars. You can imagine how Alcubierre's theory excited sci-entists and *Star Trek* fans alike. But when you look at the details of Alcubierre's theory, you realise that any celebration is premature.

The first problem is that the Alcubierre warp drive requires an energy source with what physicists call a 'negative energy density', which may or may not exist. It is entirely speculative. If it doesn't occur in nature, then the whole idea falls apart.

The next problem is that, based on Alcubierre's initial plans, the amount of energy needed to move a small spaceship across the galaxy is larger than the energy contained in the entirety of the observable universe.

Many physicists have attempted to finesse the design in order to lessen its energy requirements. Chris Van den Broeck of the Katholieke Universiteit Leuven in Belgium has calculated that if you were to shrink the size of the warp bubble so that it only contained a few atoms, then it would only take about three suns' worth of energy to power. Further work suggests that changing the shape of the warp bubble to that of a torus would reduce the energy needed even more. But this still leaves the problem of the ship being microscopic in a world where astronauts are human-sized.

These are not the only problems with the Alcubierre warp drive. Quantum mechanics suggest that extreme high temperatures inside the warp bubble, caused by Hawking radiation, would destroy anything inside it. And the ship decelerating at its destination could cause a shockwave which would destroy the destination travelled to. This is assuming that a way is found to decelerate. There does not appear to be any way in which anyone travelling in an Alcubierre warp drive ship could communicate with the cosmos outside the bubble, which means it would not be possible to steer or stop the ship once it was underway. A further problem is that the Alcubierre warp drive could, in theory, be used to travel in time, which is against the laws of physics according to Stephen Hawking's chronology protection conjecture.

If we are not able to travel faster than light, could we still reach the stars at a slower speed? The difficulties with this approach are explored in Kim Stanley Robinson's 2015 novel *Aurora*, which applies hard science to the problem of sending humans to a planet orbiting the star Tau Ceti. In the story, the spacecraft is launched in the year 2545. It is accelerated initially by being squeezed between two massive electromagnetic fields, before being 'pushed' along by a powerful laser aimed at the ship's stern. In this way it reached the speed of one tenth of the speed of light, or roughly 108 million kilometres per hour. Travelling at this pace, it can reach Tau Ceti in 170 years. Seventy-six per cent of the ship's mass is the fuel needed to slow the spaceship down sufficiently to stop at its destination. This is done by repeatedly detonating small nuclear explosions in front of the ship over a period of twenty years. All this may sound ludicrous, but current thinking suggests it is the most plausible scenario for a journey of this type.

The ship is ten kilometres long. It consists of a central spine surrounded by two ring-shaped structures, each of which is divided into 12 sections. These 12 sections contain different biospheres, each containing the flora and fauna of different Earth ecosystems, from temperate forests to tundra or grasslands. There are typically around 2,000 people on the ship, and those who will arrive at Tau Ceti are seventh-generation descendants of the crew who left Earth. They are not happy, because they didn't make the decision to journey across the stars or agree to spend their lives confined within the walls of a metal ship.

The ship is a closed system. No new resources can be added, and all energy and material must be constantly recycled. This is far from easy. Each ecosystem must be carefully monitored to prevent any loss because small ecosystems are more vulnerable to shocks or change than large ones. On Earth, the ecosystems of small islands suffer from this problem. On island ecosystems, large animals can become small and small animals become large.

The difficulties of generations of people living and dying within the ship's closed system soon becomes apparent. Each generation has a lower IQ score than their parents. The ship's bacteria are evolving at a faster rate than humans. Systems are breaking down, and the obsessive control needed to keep the ecosystems from decaying further is increasingly resented by the ship's passengers, who were born into a preordained system over which they have no say and that they cannot change.

As the character Aram explains, 'The problem is this: the spaces we have available to live in are too small to survive in [over many generations]. The main problem is the differential rates of evolution between the various orders of life confined to the space. Bacteria generally mutate at a rate far faster than larger species, and the effect of that evolution on the larger species is eventually devastating. This is one cause of the dwarfism and higher rates of extinction seen in island biogeography studies. And we are an island if there ever was one.'

When Robinson first describes this ship, he makes you marvel at the size of it. By the time he has detailed the difficulties in keeping the cycles of phosphorus, nitrogen, carbon dioxide or oxygen from being depleted, you realise how small the ship is compared to the vastness of the space it travels. The natural ecosystems needed to sustain human life are not designed for an arena this self-contained. Even an isolated island ecosystem on Earth receives inputs from beyond its shore, be that heat and energy from the sun or fish and seafood that can be caught off the coast. An interstellar ship is entirely closed. It cannot even take energy from solar panels, as it is too far away from any star.

Miraculously, the crew arrive at their new home. The gravity is too great on the massive planets around the star Tau Ceti, but a moon around one of the planets is thought to be capable of sustaining life. On arrival, it turns out to be a largely featureless rock cursed with extreme winds. It is also poisonous to humans, who sicken and die if exposed to the atmosphere. It is thought that the cause is an alien microscopic protein called a prion, which was impossible to detect from Earth. After a 170-year journey, the moon turned out to be uninhabitable by humans.

As the character Euan says, trapped on the inhospitable moon, 'What's funny is anyone thinking it would work in the first place. I mean it's obvious any new place is either going to be alive or dead. If it's alive it's going to be poisonous, if it's dead you're going to have to work it up from scratch [. . .] So what's the point? Why do it at all? Why not be content with what you've got?'

The message of the book is put most succinctly by Aram, who is called upon to address a conference about starships back on Earth. 'No starship voyage will work,' he declares. 'There are eco-logical, biological, sociological and psychological problems that can never be solved to make this idea work. The physical problems of propulsion have captured your fancy, and perhaps these problems can be solved, but they are the easy ones. The biological problems cannot be solved [. . .] The bottom line is the biomes you can propel at the speeds needed to cross such great distances are too small to hold viable ecologies. The distances between here and any truly habitable planets are too great.'

When Roddenberry dreamt up *Star Trek*, we did not know for sure that there were alien worlds orbiting the stars we see at night. In the twenty-first century, our technology has confirmed that the galaxy is teeming with planets. Kepler 62f is only one of a count-less number of mysterious and unknown worlds just beyond our reach. But as our telescopes and astronomy knowledge increased, so did our understanding of biology. As we found the alien worlds we dreamt of, we also discovered that we could never settle them.

There are science fiction writers who have attempted to dream up ways around the hard science that Robinson based *Aurora* on. A good example of this is what the American writer Kameron Hurley describes as her 'lesbians-in-space organic worldship space opera' *The Stars Are Legion*. In Hurley's novel, the starships themselves are living, breathing organisms called 'worldships'. They travel across interstellar space in packs, and they need to eat, die and reproduce. There are people on these ships, but they are just part of the larger ecosystem rather than something special or privileged. Hurley has recognised that humanity can only exist as part of a larger ecosystem, not independent of it, and she has projected this insight into the future in order to challenge Robinson's pragmatic realism. It is, however, the far, far future. No one claims that technology like living, reproducing worldships will be developed for some time. If the *Star Trek* franchise wanted to move their fictional future history further ahead in time, then hiring writers with the vision of Hurley would make this possible. But in the real world there does not seem to be a route, over the next few centuries at least, that will take us to planets orbiting a different star.

If we are to leave Earth and forge a future on an alien world, then it will have to be much closer. There is really only one candidate. Mars.

7.

Elon Musk has repeatedly said that he wants to die on Mars. To this, he usually adds, 'just not on impact'. It's hard to think of any other person who could get away with this remark. From any other mouth, it would sound delusional. From Musk, you have to wonder.

Musk is attempting to use a combination of engineering and capitalism to reduce global warming, create a sustainable energy system and make mankind multi-planetary. He intends to achieve this last goal by establishing a permanent base on Mars, of around a million people, before the end of the twenty-first century. This is necessary, he argues, in order to reduce the risk of human extinction caused by environmental collapse.

Any one of those goals would be a lofty target for any individual. For one man to try to tackle them all sounds crazy. Yet Musk has a knack for pulling off the impossible. He doesn't usually achieve his goals as quickly as he claims he will, which makes it easy to portray him as a fantasist, and his single-minded dedication to his goals often rubs people up the wrong way, which makes him a target of ridicule. But what he has already achieved makes it impossible to discount him. No one else has sent their car to Mars.

There's a scene in *Star Trek: Discovery* where a spaceship captain tries to inspire a member of his crew. He asks, 'How do you want to be remembered in history? Alongside the Wright Brothers, Elon Musk, Zefram Cochrane?' This list of names is a rollcall of great pioneers, with the Wright Brothers representing aviation, Musk representing spaceflight and the fictional *Trek* character of Cochrane representing the invention of warp drive and interstellar flight. At first glance, the use of Musk in this context feels like a snub to the true genius pioneer of human spaceflight, the Soviet rocket engineer Sergei Korolev. Yet, in the decades to come, Musk could earn his place on that list.

Musk's great talent, beyond his business and engineering skills, is to disregard the prevailing consensus about what is possible. When he co-founded Tesla Motors, the car industry considered the idea of electric vehicles to be a joke. The notion that we'd be better off with cars powered by electricity rather than petrol – a polluting fuel made from our finite supply of fossil fuels – had been around for decades, but the consensus opinion was that electric cars would always lack the speed, convenience and power of combustion engines. Petrol cars were a mature technology that the public liked and understood. There were only downsides to changing over to vehicles with a limited range that took hours, rather than minutes, to refuel. Car manufacturers were under the shared delusion that electric vehicles would never be any good, and that consumers would never buy them. That the fight against climate change required us to decarbonise our transport was not their concern.

For these reasons Tesla Motors, like many electric vehicle start-ups, was routinely mocked when it was established in 2003. Fourteen years later, electric cars are now not only thought possible, but inevitable. The UK Parliament has declared that all new petrol or diesel cars will be banned after 2040 and Scotland is aiming to achieve this by 2032. Jaguar Land Rover plans to end production of petrol and diesel vehicles in 2020, and Volvo in 2019. The European Parliament has also drawn up laws to reduce car emissions and incentivise electric vehicles, because its electric-vehicle industry is far behind China's. All this happened because Musk and his team shattered the collective delusion that electric cars would never work. Musk is easy to mock because his unwise use of social media has led to lawsuits and doubts about his leadership, but can you name anyone who has made a greater contribution to maintaining our climate for future generations?

Musk did this by looking at the problem not from the usual technical angles, but from the perspective of an engineer who included consumers as part of the solution. He knew that Tesla vehicles had to be desirable. His first vehicle, the Mars-bound Tesla Roadster, was a sports car that cost $100,000 and went from 0 to 60 mph in an exhilarating 3.7 seconds. This was not what people expected from an electric vehicle, and it appealed to consumers who did not care two hoots for the environment. With the money and branding buzz that came from selling such high-end, premium vehicles, Tesla then developed the Tesla S, a luxury hatchback with a starting price of $75,000. Each successive model was then cheaper and aimed at a wider market demographic. In this way, immense engineering challenges have been tackled, battery technology has developed, and the company is finding a way to scale up production, learning as it goes, until it is in a position to sell electric vehicles that are competitively priced alongside traditional vehicles. All this has been far from easy, and there have been many delays and economic or technical glitches along the way. But the overall achievement of Tesla has been sufficient to make electric vehicles plausible. As the *Wall Street Journal* described the Tesla 3 in their July 2018 review, 'This is magnificent [...] This is what ordinary cars should be.' The major car manufacturers, in response, have gone from thinking that developing electric vehicles would be an expensive folly to thinking that acquiring the technology is vital for their future, and the only way to keep their shareholders happy.

Musk took a similarly financially savvy approach to the even more implausible arena of spaceflight. Previously, developing space rockets had only been successfully achieved through massive government funding. Musk thought that building a private space company was feasible if he applied market principles to the problem.

The cost of a rocket, as launched by the Russian or European Space Agency, varied depending on weight and other factors, but it would typically be somewhere between $100m and $260m. Musk realised that if he could drastically reduce the cost, while still appearing credible, then the entire satellite-launching industry, both commercial and government, would be his for the taking. He built a company called SpaceX which, by 2013, was successfully launching satellites for prices ranging from $55m to $61m. This alone should have been sufficient to maintain a successful and profitable company. But this wasn't Musk's goal. This was all done to further his dream of colonising Mars. Launching satellites was a means to an end.

Most of the cost of a satellite launch was spent building the rocket, which would burn up in the atmosphere when it fell back to Earth. A SpaceX Falcon 9 rocket costs $54m to build, for example. If the rocket could be reused, this would radically change the economics of leaving our atmosphere, so Musk's team built a rocket that went up but also came back down and landed. This was like trying to land an upright twelve-storey building falling from space. To the amazement of much of the rocket industry, who knew just how difficult this would be, Musk and his team succeeded. Falcon 9 rockets, or at least their first stage, which accounts for about 75 per cent of their cost, now regularly land on the floating drone ships *Just Read the Instructions* and *Of Course I Still Love You*. According to Musk, each rocket can be used up to a couple of dozen times. There are still refurbishment and fuel costs involved, but the price tag for leaving the planet has been drastically slashed. Musk, an entirely self-made private individual, has advanced spaceflight in a way that no government-funded space agency has ever done. The consensus of what was possible had crumbled under the determination of a singular individual. As the English theatre director Ken Campbell used to say, 'If it's not impossible, it's not worth doing.'

Despite all this, Musk still has a long way to go if he's going to land men and women on Mars, especially if he intends to scale up the human presence to a sustainable community. For all the celebrity and lifestyle gossip that surrounds him, Musk is not a man we should be quick to dismiss. He achieves goals that most consider impossible. Yet the reality of Mars is a bigger challenge than anything he has yet tackled.

8.

As a base for humanity, Mars has a number of things in its favour. At its closest it is only 33.9 million miles away which, in space terms, is practically on our doorstep. It should only take between six and nine months to get there, depending on the amount of fuel used and other factors. That's a long time for anyone to spend cooped up in a cramped rocket ship, and there are many psychological and physical difficulties that budding Martians would need to overcome during such a journey, but it is doable. We could get there and, judging by the success of SpaceX's Falcon 9 rockets, we should be able to land safely.

Once on Mars we would find that a Martian day is only 37 minutes longer than an Earth day and that, unlike the moon, the planet does have an atmosphere. The atmosphere is about 100 times thinner than Earth's, and 95 per cent of it is carbon dioxide, so we're not going to be able to breathe it, but at least there's something. Gravity would be noticeably different, however, being just 38 per cent of what we are used to. It would allow us to leap long distances and lift heavy weights. This would be fun at first, but in the long term it would cause biological problems. It would cause the muscles around your spinal column to waste away, along with your quadriceps, calves and buttocks. When people talk about living on Mars, the threat to our bums is rarely mentioned.

Something else that would waste away is your skeleton, which would leak calcium into your bloodstream and cause symptoms ranging from constipation to depression. Other possible symptoms caused by low gravity include reduced immunity, nausea, heart disease and impaired sleep. Much of this could be averted with a strict physical-workout regime designed to prevent you becoming sick, sad and bumless. But, ultimately, our bodies are designed to work under Earth's gravity and not that of Mars. We don't know what impact low gravity would have on a child who was born and developed on Mars, for example, and finding out would be ethically extremely problematic.

Growing crops on Mars would not be easy. It should be possible to grow crops like a form of wheat inside custom-built structures that allow us to increase the temperature and air pressure, but the lack of nitrogen on Mars will be a serious problem. For fertiliser, we will need to rely on human faeces and urine.

The Martian soil contains a form of chlorinated hydrocarbons called 'perchlorates'. These can be beneficial to microbes that could be used as part of a process to generate oxygen, but they interfere with the thyroid and are toxic to humans. Life on Mars would need to be structured around the knowledge that Martian soil is poisonous, and that even small amounts of dust need to be avoided. Not that Martian colonists would want to be exposed to the atmosphere. The average temperature on Mars is minus 63 degrees Celsius, so no one would be kicking off their moonboots and wriggling their toes in the Martian sand.

A more fundamental difficulty is created by radiation. The magnetic nature of Earth's iron core creates a bubble of magnetic protection around the Earth, called the 'magnetosphere', which protects us from the cosmic rays and radioactive particles that pour out of the sun. It also helps protect the astronauts orbiting the planet inside the International Space Station. The shape of the magnetosphere means that, when the radiation from space is particularly intense, this cosmic radiation can sometimes be seen near the poles in the form of the Northern and Southern Lights.

Unlike Earth, Mars does not have magnetic poles to protect Martians from all this radiation. Its thinner atmosphere does not help either. The problem is particularly acute when large solar flares erupt, sending great masses of particles out into space, but radiation is a constant presence for anyone outside of Earth's magnetosphere. At best, this will lead to an increased risk of cancer, and at worse to serious radiation sickness. Spacecraft, spacesuits and Martian buildings all need to be constructed with this in mind, and ways to shield their inhabitants need to be engineered in. The most effective defence is to move Martian buildings underground. It does not seem to be a coincidence that in late 2016 Musk founded the wonderfully named Boring Company, in order to develop technology that reduces the cost of underground tunnelling. 'For sure there's going to be a lot of ice mining on Mars, and mining in general to get raw materials,' he explained. 'And then, along the way, building underground habitats where you could get radiation shielding [...] you could build an entire city underground if you wanted to.'

For someone with Musk's can-do attitude, all this is just a list of problems that we will have to work on and solve. Yet at this point it is probably fair to ask, why would you do this? Musk's argument is that humanity needs to become a multi-planet species in case an environmental disaster made Earth inhospitable, but what potential environmental disaster could befall Earth that would make Mars a more hospitable home? Even if an atomic war triggered a nuclear winter, does it make sense to move over 30 million miles to somewhere that also has unsafe radiation levels and is too cold for human life?

Imagine if life on Earth became so perilous that it became necessary for a million survivors to move to the unspoiled continent of Antarctica and live in tunnels underground. This would not strike many people as an appealing way to spend the rest of their days, but it would offer a far greater quality of life than living underground on Mars. The air contains oxygen and would be breathable, the temperature is warmer, water would be easy to obtain, the ground is not full of poisonous perchlorates, sufficient quantities of nitrogen exist on the same planet, radiation isn't a problem and you would not suffer the many and varied medical conditions caused by low gravity. In comparison to Mars, living underground in Antarctica would be a life of luxury. Yet who would want to spend their life doing that?

Or, if not Antarctica, what about living at the bottom of the ocean? There is no disaster scenario that necessitates moving to Mars which could not also be solved through huge numbers of survivors living on the floor of the Pacific or Atlantic. It would be cheaper, safer and healthier, and it does not require the invention of major new technology. We would find an environment as alien as any far-off planet, teeming with life and wonder. We will not find anything as fascinating or as ornery as an octopus on Mars. But we would long for life on the surface after a number of months had passed, when the realities of our dependence on pressurised suits and habitats had become frustrating and the novelty of life on the seabed had begun to wear off. If we don't want to live permanently underneath Antarctica or the oceans, why would we choose the vastly more dangerous and expensive option of living a worse life on Mars? Yes, the dream of living on Mars is cool, or at least it is while nobody has done it. But it is the reality of living on Mars that matters.

Some believe that one day we will be able to terraform Mars. This would entail artificially altering the Martian atmosphere until it is made up of an oxygen-rich, breathable atmosphere that could trap sufficient heat to give Mars a similar temperature to Earth. How this can be achieved is unclear, because there does not appear to be enough carbon dioxide on Mars to sufficiently thicken and heat the atmosphere. If a solution is found, Mars would still be bombarded with cosmic radiation, riddled with perchlorates and lacking in gravity, but it would be more hospitable than the Mars we know now. Against this dream, it is worth considering the difficulty we have keeping our own atmosphere stable, as temperatures increase and runaway climate change seems dangerously likely. Fine-tuning an existing atmosphere is extremely difficult given the complex nature of the many differing feedback loops involved, but it is nowhere near as difficult as building a complex biosphere from scratch.

You could argue that the real danger is that Earth will one day become overcrowded, and that is why we need to go to Mars. If that is the case, the sensible approach would be to tackle overpopulation now by promoting female education and access to contraception in poorer countries. This would be far cheaper, easier and less dangerous.

This is not to say that no one will go to Mars, or build a moonbase, or mine nearby asteroids. On the contrary, I think those things are almost certain, and I hope I'm around to see them. Such adventures would give us a great deal of new scientific knowledge, and there will be brave souls prepared to suffer in order to do this. But it does not logically follow that mankind will then leave Earth in sizeable numbers and found sustainable communities on other planets.

It seems more likely that it will be robots, not humans, that will journey to space in great numbers to swarm around the solar system and mine the asteroid belt for resources. Such activities, it is looking increasingly likely, will be run and financed by the private sector, and the economics of the situation massively favours robots over humans. A robot doesn't need to breathe, eat or be kept at human temperatures and pressures. If it costs 100 times more to send a human to mine the moon than it does a robot, then don't be surprised when those private corporations choose robots.

Humans evolved alongside a symphony of flora and fauna in the thin damp band of atmosphere that clings to this planet. It is the only known place in the universe where we can survive. We can live only within a tiny range of temperature, pressure and gravity, and we need a diverse range of living things, from bacteria to plants and animals, in order to do so. We can create bubbles of this ecosystem, such as spaceships and spacesuits, that will allow us to survive out in space for a while. But unless we can build another Earth, or alter ourselves so severely that we are no longer recognisably human, then these bubbles will be too small to sustain us indefinitely. We are not self-contained individuals, ready to step away from home to forge a new life out in the cosmos. Humanity is part of planet Earth's ecosystem, and we will die if taken outside of it.

9.

As someone raised in the late twentieth century, the realisation that our future is not out in space has been something of a shock. I remember, as a child in the late 1970s, having a *Star Wars* poster on my wall. It showed Luke Skywalker outside his family farm on a dead-end desert planet, looking out at the horizon and the setting of twin suns, dreaming about leaving his planet to find adventure in the vast galaxy beyond. When I was a little older I spent countless hours playing the videogame *Elite*, flying a Cobra Mark III spaceship between space stations across the galaxy. The *Star Trek* myth that our future would be out in space was always present. I cannot recall hearing voices in our wider culture that said we would not leave the planet, and that we must always live on Earth.

This realisation brings mixed feelings as you move your eyes down from the heavens and look at the green and familiar world around you with fresh eyes. As the idea sinks in, you realise that this is really not a bad situation to find ourselves in. There is nothing in the known universe more complex, beautiful and awe-inspiring than our living planet. There is nothing as rich or as intricate, and there's nowhere we would be better off. We are not here as tourists. We're as much a part of this unfolding biosphere as the rest of life on Earth.

We are going to have to make this work. We're going to have to remain here sustainably. When we think of the future, we need to think of the human world, the natural world and the digital world as being co-dependent and predominantly earthbound. All three live on this rock, and we will all have to get along.

There is a type of story we can use to help understand this. It's a structure where a small bunch of wildly different characters are trapped by fate into occupying the same space. They dream of escape, but that escape can never arrive. This story structure is the sitcom. The 'sit' refers to the situation, often a family home or workplace, which fate has trapped the characters in. The 'com' refers to the comedy that naturally arises as those very different people try to get along.

What's interesting in this analogy is that a sitcom is sustainable. For all the wild antics that occur in each story, the equilibrium always resets itself by the end of the episode. It is not like a drama, where events change the dynamic of the setup so that things are never the same again. A well-structured sitcom can run for years and years, if not indefinitely.

Sitcom, then, is the best metaphor for our future. Humanity, our digital creations and mother nature attempt to get along, while trapped together on the third rock from the sun for untold years to come. Would that be so bad? Characters in sitcoms often dream of escape from their situations, yet the watching audience love to return to that world. They get attached to the characters and want to spend more time with them. They look forward to the next episode because they see the humour of the world. They know wild things will happen, but they are allowed to enjoy such things because they trust the world to return to its stable dynamic at the end of the episode.

If sitcom characters realised that they were in a sitcom, could they enjoy their lives on the same level that the watching audience does? It would take a shift in perspective similar to seeing the patterns in big data as well as the flaws in individual details. A metamodern generation could, perhaps, be a character in a sitcom and enjoy that sitcom at the same time.

And just because outer space is not for us, it does not mean that there aren't great universes for us still to explore. We just won't find them out beyond the horizon. Instead, we're in the process of dreaming them up.

6.
BETWEEN THE REAL
AND THE VIRTUAL

1.

I am standing in a white void. Or, at least, subjectively I am standing in a white void. Objectively I am standing inside the Netherlands Film Academy in Amsterdam, wearing a virtual-reality (VR) headset. But, from my perspective, I'm in a white void populated by a number of floating spheres, about a metre across, which I walk around and peer at. Distorted, unrecognisable video footage flickers across the surface of each sphere. I've been told to walk up to each sphere and just poke my head inside.

I choose a sphere at random and walk up to it. It hangs in the air, at roughly chest height. I take a breath, bend down and push my head inside. The moment I do this, I find myself in a city park somewhere. I am standing in the grass, and I am only an inch or two high. Giant people run up and jump over me, which is nothing if not disconcerting. Then I pull my head out of the sphere and find myself back in the white void. Choosing another sphere, I enter and find myself in the centre of a roundabout in a children's playground. This is then spun around at speed, and as a result I nearly fall over. In a third sphere, a man picks me up and places me in his mouth. That was even less fun than the roundabout.

Removing the VR helmet, I see that two of the people who appeared in the spheres are with me in Amsterdam. One is the filmmaker Nik Alderton, who was the person who had just placed me inside his mouth. He doesn't immediately try this again in the real world, but I keep one eye on him just in case. Instead, I go and talk to the London-based writer Michelle Olley.

Michelle has an uncanny ability to find herself in vibrant and radical artistic communities, seemingly without trying. One example was Voss, probably the most celebrated show by the troubled but visionary fashion designer Alexander McQueen. This featured a large brown box at the centre of the stage. At the end of the show, the sides of this box fell away to reveal a large naked woman, covered in live moths, with her face hidden by a gas mask. That was Michelle. The show featured models like Kate Moss and Erin O'Connor, but it is Michelle that everyone remembers. Given her innate ability to stumble into fascinating but unconventional circles, the fact that she is exploring VR strikes me as a reason to pay attention to it. I ask her how she made the void and the spheres.

'They were done using 360-degree videos,' Michelle says. 'Me and Nik made them a few months back on one of Scott's training courses.' Scott is the designer Scott McPherson, who is responsible for bringing me to Amsterdam. 'Nik and I were partnered up on the day, and Scott sent teams of us out into Brighton and said, "Go and make some films." The idea was to start thinking in terms of your complete surroundings, everything around you, not just that oblong frame that you concentrate on with normal cameras. So, we all went out with a 360 video camera, which is just a tiny little thing that films in all directions at once, and we all took bits of video in the park and in corner shops. We did things like, we bought a cake and fed it to some pigeons and filmed them going round the camera pecking at crumbs and stuff. Yeah, it was quite playful. Or we took the 360 video camera and stuck it in the cold-drink cabinet in a shop, switched it on, and reached in to take drinks. The result was these giant hands coming towards you like something from Monty Python. Basically, we were just messing about to see what this new tech can offer.

'What I like about that white-void demo is that it is not what you expect from VR. You think of VR as going from the real world to a computer-created world, which is initially what happens, but then when you put your head into a bubble you go from a computer space to the real world, or at least the appearance of the real world, when you're surrounded by 360-degree video. The whole thing's topsy-turvy. I guess if you wanted people to move from one story to another it would be quite a nice storytelling device.'

'You make it sound easy,' I say.

'That was the point of the class really. The technology is already at an entry level where you can go and just do something and make something. It's really accessible. There were a variety of people from different backgrounds on the course – there was a girl from theatre, a photographer, a fashion designer and myself. My background is journalism and TV, I'm not a coder or technical or anything. As well as 360-degree video, making virtual-reality environments in a computer is also getting to the point where anyone can do it. We were using free software called Unity and it's not that much harder to build a VR world than it is to build something in Minecraft. There's a library of models that you can add into your world that are off the shelf and free. You can put mountains in there, you can put buildings in there. You can build things and make them yourself or you can buy a few more bespoke objects, like Scott bought me a corgi. What he was teaching us was how accessible VR actually is. It's not just about passively experiencing it. You can make your own VR environments. There are tons of kids doing it.

'This was how that spheres demo came about. Scott made a VR environment that was just a few spheres and nothing more, then he pasted the videos we shot on the inside of the spheres and set it so that they would start playing as soon as you moved inside them. It took, what, five minutes at most to build? It's the old punk idea of "Here's a chord, here's another chord, here's a third, now go form a band." Now obviously, the level of Unity knowledge that he used to create the bubbles and put them within that environment is higher than mine, but it's not that much higher really. It wouldn't take long to figure it out. The idea that VR should be something you create and not just experience isn't something that's really talked about. But it is sort of happening without it being talked about because kids are getting into it.'

Among the junk and goodies I found in the back of a cupboard during a recent clear-out was a pile of early 1980s *Load Runner* comics. This title billed itself as 'The Galaxy's First Computer Comic'. It contained a bunch of comic strips about characters trapped inside computers and forced to play games for survival. It reminded me that home computers were originally thought of as portals that you entered. The film *Tron* is an obvious example. Timothy Leary used to refer to the experience of using computers, and moving your mind from the real world and into the virtual, as 'passing through the Alice window', a reference to *Alice Through the Looking Glass*.

When the comic was written, technology wasn't advanced enough for us to really feel that we were entering the computer world. It came a step closer during the 1990s, thanks in part to the work of the dreadlocked American computer scientist Jaron Lanier. Lanier established VPL Research, the first company to sell VR goggles and gloves, and coined the name 'virtual reality' to describe this technology. Unfortunately the processing power needed to do justice to his designs was then unavailable, or unaffordable. It has taken until the second decade of the twenty-first century for graphics chips to get fast and cheap enough for home VR technology to become possible. As a result there has been an upsurge of interest in the technology, and VR headsets are now available for PlayStation and PC games. The initial promise of games, that they were worlds we could step into, is coming to fruition. But there are reasons to be cautious.

One idea that has been prevalent in recent interest in VR is the notion that the technology is an 'empathy machine'. VR allows you to step into the shoes of others, and to experience their world as they see it. The idea that such technology can create empathy for far-off people chimes nicely with Generation Z culture, and as a result it has received a lot of attention. But back in the early days of virtual-reality development, benefits like these were usually balanced with concern about other, darker aspects of the experience.

Back in the 1990s, Jaron Lanier was deeply troubled by the idea that VR could be used to control people. 'A suffocating thought hit me only a few months after my initial computer graphics reveries,' he wrote. 'It was so awful that I had to unthink it right away; it burned. [...] The thought is that virtual-reality technology is inherently the ideal technology for the ultimate Skinner box. A virtual world could be, precisely, the creepiest technology ever.'

A Skinner box was a device invented in the years following the Second World War by the American psychologist B. F. Skinner. At its simplest, it was a box big enough to hold a pigeon, rat or other small animal, and which contained a food dispenser and either a button to peck at or a lever to pull. More complicated versions of Skinner boxes could also include lights, speakers and an electrified floor capable of hurting the animal inside, but the simplest version is sufficient to demonstrate something quite disturbing. A Skinner box showed that it was possible to give an animal the illusion that it was in control, even though the animal's behaviour was being controlled by the owner of the box.

Achieving this was more complicated than just giving the animal a morsel of food every time a button was pressed. Once a pigeon had realised the connection between the action and the reward, it would repeatedly peck at the button in order to trigger deliveries of food. However, these actions would not continue indefinitely. After a certain period, the pigeon would be full, and lose interest in this perpetual feast.

If you wanted to make the pigeon keep pecking, Skinner realised, you had to override the pigeon's regular behaviour. You could do this if you were crafty about how you linked the action to the reward and did not provide food every time the button was pressed. Instead, a button press only released food sporadically. Moving from a guaranteed reward to the possibility of an occasional reward, he discovered, made the pigeon keep pecking away. Like a gambler at a slot machine, the pigeon believed that it was continually pushing the button out of its own free will. It was the design of the box itself that was responsible for the pigeon's repeated actions, but the pigeon was blissfully unaware of this.

For Lanier, it was obvious that Skinner boxes were morally indefensible. The concept is horrifying, which is why the reveal in the movie *The Matrix*, that humanity is trapped inside highly sophisticated VR that functions like a Skinner box, is so terrifying.

Lanier's words remind me of warnings about VR that I've heard from Scott McPherson, the designer who trained Michelle. Scott's office is on the top floor of a tall Georgian terraced building near the centre of Brighton. I've climbed the stairs to that room many times over the last five years, and each time the VR he's shown me has advanced in leaps and bounds. A few years ago, VR used to make me feel nauseous after only a short demonstration. Now, it's become a portal to worlds where you can relax, explore and spend hours in. Like Lanier, Scott can make you giddy with excitement about the potential of VR, but, also like Lanier, he has words of warning about it. After returning home from Amsterdam, I decide to pay him a visit.

2.

Scott, I should point out at the start, is a spiky one. He is the only person I know who has spent a night in jail after an argument about a font turned violent. He does not play nicely with others and he works under the name 'amoeba': a cell of one. Bearded, tattooed and never still, Scott sports a ginger quiff as simple but as meticulously crafted as Soviet propaganda poster design. I ask him what it is about VR that concerns him.

'Control,' he tells me. 'Aye, that's the thing. VR is revolutionary and brilliant and I love it, but it's power, and bad bastards can use that power as easily as the good guys.

'I learnt about control in the nineties, when I was doing all the visuals at clubs. When you knew what you were doing, when you knew the tricks, you could control four hundred people in a nightclub. I could make them happy, sad, stop them dancing, start them dancing – just by a push of a button. I'd control them through typography, animations, the speed of animations, the colour changes, flashes, all sorts of stuff. When you know those techniques, you can take a whole audience where you want and you become a puppeteer. We're more open and susceptible to other people's influences than we think.

'What I knew then, as a VJ and an audiovisual guy back in the nineties, was that people could be made to believe things and they were susceptible to manipulative mind tricks. Images and visuals have an effect. People will be influenced by these things even though they think they won't. You're putting things in people's heads and you have to take responsibility. You have to, because we can now affect people's happiness and sadness, and that's a big responsibility. And if it was me, a working-class guy from Paisley in Scotland, and I could do that to four hundred people, imagine what a company with resources could do? News organisations and politicians, TV companies, Hollywood – imagine the influence they actually have on people's realities and brains.'

Scott gets up to make a small, insanely strong cup of coffee. 'How does that work?' I ask.

'You don't need to imagine it, you can see it now, the reality of it. It's how for the last three years, with Trump and the Russians and fake news and Brexit and all the idiocy, they got us and we never realised it was happening. Do you think millions of people just woke up one day and decided to become idiots? Of course not. They were made into idiots as they sat in front of a screen while their brains were being rewired. When you come to things passive, that's how you get done.

'When you're passive these things can be imprinted on you. You're just sat there at the screen and you're slumped a bit and you're not fighting it, you're just letting it imprint on you and then they've got you. TV people have known that for ever, advertisers have known that for ever, everybody's known that. Whereas, if you're using your imagination, then you're not passive and they can't get you. People like [the writer] Alan Moore know this. They leave a space in their work for your imagination, and that way you're not being controlled, you're being empowered. But not many people do that.

'Look at [the misogynistic cultural movement] Gamergate and all that. All these dumb wee boys were done good and proper.

They didn't know people were using things against them and binding them to other realities and putting them in a reality tunnel that wasn't theirs. They were radicalised in exactly the same way that Islamic fundamentalist boys were radicalised – exactly the same way. And in part that was done by bad bastards who did so deliberately, and in part by algorithms who had no idea what they were doing, but they had been let loose and allowed to play a part.'

What Scott is referring to here is deliberate recruitment of young people, especially boys, to far-right and white-supremacist movements. Much of this is done online in gaming communities using text chat apps such as Discord. A common tactic is to first interest targeted people in non-political conspiracy theories, such as UFO cover-ups, because after someone has begun to question one part of societal consensus it becomes easier for them to consider other more extreme ideas. This process is known as 'redpilling', after a scene in *The Matrix* in which Keanu Reeves's character is offered the choice of a red pill or a blue pill. The blue pill leaves you safe and secure in the normal world, while the red pill is said to reveal how the world really works.

The far-right notion of being redpilled is similar to the liberal notion of becoming 'woke'. Both terms express a belief in becoming enlightened about, and transcending, the flawed unconscious prejudices of normal society. Both notions are extremely tribal. To insiders, this identity is justified as promoting increased empathy and support for other members of the tribe, although it can produce antipathy for outsiders. To those outside the tribe, it can appear that this contempt for others is the driving force of the movement and increased in-group empathy just an excuse.

The main difference between being redpilled and becoming woke is the issue of power, and where that power is located. A person who considers themselves woke will identify with sections of society who lack power, and who experience prejudice and dismissal in the wider culture. Someone who considers themselves redpilled, in contrast, will identify with a powerful privileged majority, to the extent that they will need to create imaginary or exaggerated enemies to threaten them. Those enemies may be nebulous concepts such as 'cultural Marxism', which is a term that makes very little sense to people outside far-right tribes, or dehumanised caricatures of foreign people or religions.

This process of online tribal recruitment is complicated by the use of bots to promote antagonism between the tribes. According to a University of Southern California study, the majority of online discussion about the 2017 Disney movie *Star Wars: The Last Jedi* was by 'bots, trolls/sockpuppets or political activists using the debate to propagate political messages supporting extreme right-wing causes and the discrimination of gender, race or sexuality. A number of these users appear to be Russian trolls.' That Russian forces are attempting to destabilise Western society by moaning about *Star Wars* seems bizarre, yet it illustrates how effectively the antipathy between tribes can be manipulated to produce a fractured, weakened society.

Some traditionalist *Star Wars* fans didn't like the movie because the story argued for letting go of the past in order to move forward, which was a brave theme for a nostalgia-focused franchise. It is doubtful that when those fans discussed their opinions online they realised their views would become an excuse for far-right extremists to abuse non-white, non-male *Star Wars* actors such as Kelly Marie Tran, who was on the receiving end of such an outpouring of hate that she quit social media. They almost certainly did not think they were aiding foreign attempts to destabilise Western society. But, as the shadowy figures that run Russian troll farms understand, tribal cultural and political affiliations are so interrelated and complex that exploiting them to cause anger and discord can have powerful results politically.

'That's all just done with the internet, that's not using the VR tools yet,' Scott continues. 'But with VR all that will be much easier and much more powerful. People talk about VR as an "empathy machine" but it's not an empathy machine, it's an imprinting machine. I know how these things work – it's easy, I could make you feel empathy for a carrot if I wanted to. But it's not empathy if it's imposed on you. It's imprinting. And once you know how imprinting works you can make people believe anything. News in VR is very, very dangerous. Very dangerous. It can imprint you on a reality where you think, "That's it, that's the reality and those people are the enemy", when that's just not the truth. That's why news should have no place in VR.'

Scott's words remind me of the difference between two visions of dystopia, Pixar's animated 2008 movie *WALL-E* and Ernest Cline's 2011 novel *Ready Player One*. In *WALL-E*, future humans spend their days in front of screens, passively consuming the entertainment provided for them. They have evolved into obese adult children, who can barely walk for themselves and rarely leave their comfortable floating chairs. *Ready Player One* depicts a world of economic collapse devoid of real human connection, but the passive screens of *WALL-E* have been replaced by active engagement with the world of VR. In doing so, the characters have been given a sense of agency. Although the protagonists are kids, their active nature makes them seem less like children than the passive adults of *WALL-E*.

In the book *Ready Player One*, and the subsequent film adaptation by Steven Spielberg, the plot does not concern itself with making the actual physical world a better place. Instead, the story revolves around an attempt by a multinational corporation to gain control of the virtual world. Preventing the virtual world from becoming a tool for control is presented as being more important than solving the problems of the physical world. On first reading, this felt like a dangerous misstep, because surely solving real-world problems is more important than fixing make-believe? But after speaking to Scott I start to wonder whether Ernest Cline was on to something.

In the world of VR, the issue of control inevitably brings up a company called Oculus VR. Oculus was embraced by the gaming community when it was founded in July 2012, because gamers were excited by its promise of developing affordable VR headsets for videogames. It launched a crowdfunding campaign on the website Kickstarter and hoped to raise $250,000 for the development of a head-mounted display. The campaign quickly raised nearly ten times that amount.

In March 2014 Oculus was bought by Facebook for £2bn. Facebook explained that it would be using the technology to explore the social potential of VR. The community that supported Oculus did not take this announcement well. They were not happy that the focus of the company seemed to be moving away from gaming. The realisation that the money they had contributed in good faith to the crowdfund had, in actuality, provided free R&D for a multibillion-dollar corporation did not help matters.

Neither did the actions or politics of Oculus's co-founder, Palmer Luckey. Luckey, who had become the public face of VR during the rise of Oculus, became incredibly rich through the sale of his company. *Forbes* magazine estimated his new fortune at $730m and placed him at number 22 on its 2016 list of the richest American entrepreneurs under 40. He used some of this money to further his political views. Luckey is a Libertarian and his most notorious gesture of political support was to give $10,000 to the pro-Donald Trump group Nimble America, who funded anti-Hillary Clinton billboards and attempted to bring about political change through 'shitposting' and 'meme magic'. In meme magic, certain visual images are understood to have the power to affect change in the wider world when they are placed in the minds of others. During the 2016 election, the spreading of images of a cartoon frog called Pepe was understood by some to be a magical act to make Donald Trump president. This occurred at a time when, according to the pundits, political class and pollsters, there was no chance at all that Trump would be elected president.

The reaction to the news that a senior Facebook multimillionaire was funding the practice of magic for political purposes caused a sort of bemused semi-scandal. Political commentators were doing their best to grasp the strange changes that were then washing over the political world, but meme magic was too strange for most and a development they didn't even pretend to understand. The key to what was happening was that Luckey came from the world of VR rather than social media. At first, Mark Zuckerberg did not seem to understand the role Facebook played in the 2016 American election, saying that the idea that fake news on the platform had an impact was a 'pretty crazy idea'. He has since accepted that Facebook does have a case to answer, but he may have grasped this sooner if his background was in VR rather than social media.

Luckey left Facebook in March 2017. Neither party issued a statement to explain this parting of the ways, but Luckey's now-toxic reputation is widely assumed to have played a role in this split. Even after Luckey's departure, Facebook and Oculus are still regarded with suspicion by the VR community.

'What's happening now is that Oculus and Facebook are using a "walled garden" approach to the content that is allowed on their headset, which is the same as what Apple did with the iPhone,' Scott tells me. 'They control what is allowed on it and that protects them from hackers and various attack vectors and all sorts of stuff. This gave Apple control of the medium and allowed them to take porn out of the Apple environment. Steve Jobs never wanted that. I agree with him on that. He deliberately made it hard for porn to get into the Apple walled garden ecosystem by not allowing Flash videos and things like that. That's what a walled garden can do, it can allow you to stop things you don't want, and that's fine.

'But the VR walled environment for Facebook and Oculus is all about monetising, advertising, tracking, control, information analytics, the whole thing. Wherever you are in that virtual space and time, wherever you're looking, what you're doing, who you were looking at, who you were involved with, who you were online with, your speech patterns, what time of day it was – all of that is marketable information that they can sell on to data analytic companies who can then put you in a category and understand how you're going to behave, understand your beliefs and inter- actions, and then they can market to you perfectly with a pop-up advert in VR that's tailored to you alone. That's the Facebook business model. That's exactly the Skinner box that Jaron Lanier warned us all about.'

The idea that Oculus will store all possible data and give it to Facebook is a common assumption in VR circles, especially after Facebook's vice president of ads and business platform, Andrew 'Boz' Bosworth, was put in charge of Oculus. The Oculus privacy policy explicitly gives it permission to do this, declaring that 'We may share information within the family of related companies that are legally part of the same group of companies that Oculus is part of, or that become part of that group, such as Facebook.'

'Imagine being in that world – all these pop-up boxes follow- ing you around wherever you go asking, do you like this? You'll love that!' continues Scott. 'That's pure Charlie Brooker. That's *Black Mirror* land, that's not for me. You can see that's Oculus and Facebook's goal – that's obvious, completely obvious. But imagine having a news system controlled by politically active people like Rupert Murdoch in Oculus. Do you want them delivering the news in a medium where imprinting is so easy? That's a no-no. It's a nest of worms. Oculus and Facebook want that, but that's Babylon.

'Luckily, the rest of the industry are not going that way. They're going for an open platform. That gives me hope. That's a future I like. The Oculus side of it – not so much, but open-platform VR is something to work for. I'm wary because I've been here before. We had the same dream with the internet back in the nineties. We thought it was ours because we had the tools. I was well into hacking back then, but at no point did I think that hacking would be a nation state warfare tool. I thought it was still going to be underground, and that hackers were cool and against the system and all that. At no point did I think that hackers were going to sell out for the money. At no point did I think that hacking would be government policy. I saw the Utopia of the internet, like most people did, in the late eighties and early nineties. It's still the 1990s inside my head. We'll keep fighting to keep the bastards at bay. But we know them now. We know how to spot them. It's when they go for control that they give themselves away.'

This reminds me of something that Michelle told me. 'What interests me about VR is what it can show us about what we normally call reality, and about how we trick ourselves and fool ourselves into believing arbitrary things. When we made those bubbles that you could put your head in, we were talking about them as reality tunnels, and talking about using VR to make it easy to change reality tunnels. Often in VR, you are triggered to feel empathy or fear. When you realise how these things work in VR, you see how they work in the real world too. VR can give you a really visceral epiphany about these things.

'When you start thinking about creating VR you keep coming back to those two things, shock and empathy, because most VR experiences and narratives that are being created now are all about tapping into shock and empathy. You have to be careful about this because in the wrong hands this is technology that can be used to manipulate feelings. But the eternal optimist in me is kind of hoping that the result of this will not just be more malleable people, but that it could actually create more self-awareness. In the hands of the digital native generation, I think it could help them understand when their strings are being pulled.

'I know that's not going to be true for everybody, and I hope VR doesn't become some terrible tool for the dark side. Pulling on people's fears instead of people's hopes would be awful. This is why we have to engage with the tech ourselves and get it in the hands of the artists and not leave it to the corporations. VR is such a vivid experience that it feels like it matters, and if it makes a few people realise how many of their responses are automatic, then it's worth pursuing.'

Michelle's words echo the thoughts of Jaron Lanier, who is hopeful that the problems of social media will prevent us from making bigger mistakes in VR. As he remarked in a 2018 interview, 'As bad as [the problems with social media have] been, as bad as the election interference and the fomenting of ethnic warfare, and the empowering of neo-Nazis, and the bullying – as bad as all of that has been, we might remember ourselves as having been fortunate that it happened when the technology was really just little slabs we carried around in our pockets that we could look at and that could talk to us, or little speakers we could talk to. It wasn't yet a whole simulated reality that we could inhabit. Because that will be so much more intense, and that has so much more potential for behaviour modification, and fooling people, and controlling people. So things potentially could get a lot worse, and hopefully they'll get better as a result of our experiences during this era.'

'I think there's a kind of collective will among this next generation, the digital natives, to make their own narrative and not to depend passively on ones from the past,' Michelle told me. 'I think VR will probably have a part in that. The next generation will embrace it in a way that we won't – and by "we", I mean Xers, Boomers and to an extent Millennials as well. VR is not fully developed yet, it's not complete. If I was to use an analogy I think that if current VR was social media, then VR is currently Friendster and we're moving into VR as MySpace, but we haven't got to VR as Facebook yet. That's how I see it in my head. But it's getting there. There may be some steps backwards, and it might not be as soon as some people hope, but VR should become part of our lives.'

3.

Returning home, I open my laptop and find that a new VR feature has appeared on the *Guardian* website. Ignoring Scott's advice, I strap a VR headset to my face and take a look. It is a video report about the ongoing impact of the 2011 Fukushima nuclear disaster. In the video, members of Greenpeace escort former residents of Namie, a still-evacuated town north of Fukushima, back into the radioactive zone to see where they used to live. It was filmed in 360-degree video, allowing me to look around the disturbingly quiet town as if Greenpeace were taking me along with the ex-residents. The streets are empty, and shrubs and weeds are starting to take root in cracks in the concrete pavements. When these former residents tell their story, the VR makes it look as if they are standing just in front of me, looking me in the eye. It is hard not to empathise with them or wonder what it would be like to be forced to abandon your own home following some unthinkable technological disaster. This is powerful and moving, so you can see why the *Guardian* is utilising this technology. Although no arguments for or against nuclear power are given, I have little doubt that Greenpeace produced this film because of the negative impact it would have on people's views about nuclear energy.

But nuclear power is a complicated subject, and an eerie VR experience of empty streets is not the best way to judge it. Despite its scary reputation, nuclear is one of the safest forms of energy generation we have. Given the tone of the news coverage of the Fukushima disaster, it surprised me to learn that not a single person was killed. As the Harvard scientist Steven Pinker points out, large numbers of people are regularly killed in the generation of energy from fossil fuels, but this is not considered newsworthy. 'Compared with nuclear power, natural gas kills 38 times as many people per kilowatt-hour of electricity generated,' Pinker has written, 'biomass 63 times as many, petroleum 243 times as many, and coal 387 times as many – perhaps a million deaths a year.' Like air travel, nuclear power is incredibly safe because the absolute necessity of safety is ingrained in the minds of the engineers involved. After the Fukushima accident, the environmentalist George Monbiot wrote that 'the events in Japan have changed my view of nuclear power. You will be surprised to hear how they have changed it. As a result of the disaster at Fukushima, I am no longer nuclear-neutral. I now support the technology.'

Monbiot now supports nuclear because it is not plausible to move to a decarbonised energy system in the near future without including nuclear power in the mix. The advocacy group the Union of Concerned Scientists also dropped its opposition to nuclear power in November 2018 for the same reason. If public distrust of nuclear prevents the construction of new nuclear power plants, more fossil fuel will be burned instead and the chances of preventing runaway climate change grow smaller. Fear of nuclear testing was a key reason for the founding of Greenpeace in the early 1970s, and anti-nuclear policies are deeply ingrained in their organisation. But while problems such as the storage of nuclear waste may have been valid reasons to protest against the technology back in the 1970s, they are now of secondary concern compared to the damage climate change will do.

All this makes the Fukushima VR video a good illustration of why news should have no place in VR. The video seemed to be fair and uncontroversial, but it was potent propaganda. It had none of the context and information you would need to form an opinion on nuclear power, yet it created an association between the technology and a feeling of eeriness and unease. The video offers an emotional reaction instead of informed understanding. You could watch it and in no way feel manipulated, but if you were then asked your opinion on nuclear power a week later, I suspect you would give a more negative response than someone who had not been exposed to the video. If you think about how partisan and biased the press can be, would you trust them to use technology like this ethically?

Of course, all new media arrive with claims that they will usher in a moral downfall. Socrates feared that using the written word would 'create forgetfulness in the learners' souls, because they will not use their memories'. Traditionalist condemnation greeted the novel, radio, television, videogames and the internet. In all of these cases, use of these media quickly became accepted despite those criticisms, because their negative qualities were deemed to be outweighed by their positive ones. Being wary of such technology is understandable at a time when we are still coming to terms with the extent to which fake news, Russian troll farms and psychological propaganda have affected our culture and the democratic process. In time, we will become more familiar with VR and how it affects us.

The VR visit to Fukushima was fascinating in a way that depictions of that abandoned landscape in other media can't match. If it had been included as part of a wider report on nuclear power, which did include context and analysis, then it would have been a valuable contribution to public understanding. If there is one thing that a metamodern culture is good at, it is utilising the strengths of something while keeping an eye on its faults. We no longer approach an item of content simply thinking, 'What is this?' Instead, we are learning to automatically think, 'What is this, who has done it and why have they done it?'

Ultimately, this is a question of trust. VR is a potentially very powerful medium, but if you wish to use it you will increasingly need to gain the trust of your audience first.

4.

As Michelle pointed out, for all that VR has come along in leaps and bounds in recent years, it still has a long way to go. Consumer headsets such as the HTC Vive and PlayStation VR are now available to buy, but the experience they provide can be surprisingly passive. Your role is often to stand in silence as things happen around you. Much of the video available for such headsets consists of virtual rollercoaster rides, worthy explorations of historic sites and oddly submissive porn. VR games are more interactive, but they tend to be limited by simple controls using very few buttons. In VR, you can't look down from the game to check where your fingers are on the keyboard or games controller. All this acts against the fully immersive experience that virtual reality aims for.

It should not be this way for too much longer. The development of haptic gloves is making great strides and looks set to revolutionise our experience of virtual worlds. 'Haptic' refers to the sensation of touch, and a haptic glove allows you to touch, feel and manipulate objects in VR. Prototypes are so sophisticated that, should a virtual spider crawl across your hand, you will be able to feel every one of its legs as it moves. The gloves can also restrict the movement of your fingers so that if you were to pick up a virtual rock and squeeze, you would feel it in your hand and not be able to put your fingers through it. This ability to manipulate the virtual world does wonders for the brain's ability to accept the virtual world as real, and it massively opens up what you can do in virtual space. Consumer versions of haptic gloves and, eventually, full-body haptic suits, should be with us soon, as will headsets that track where your eye is looking and adjust the focus and detail of what you see accordingly. At that point, the old idea of the digital world being a portal through which you enter will, to all intents and purposes, become a reality.

We are only starting to discover the potential uses of this technology. One interesting area to be explored is psychological therapy. In an experiment conducted by a team at the University of Barcelona, subjects wearing a full VR bodysuit were given the experience of being inside Albert Einstein's body. When they looked in a virtual mirror, they saw Einstein looking back at them mirroring their own moves exactly. Tests performed before and after the experiment showed that becoming Einstein made people less likely to unconsciously stereotype older people, and people with low self-esteem also scored better on a cognitive test. VR therapy could give us the ability to question our unconscious assumptions about ourselves and others. It can also help us overcome our phobias. In an experiment at Stockholm University, adults with public-speaking anxiety used VR to conquer their fears, and reported feeling less nervous, shaky and sweaty, and significantly more confident, after practising public speaking in the virtual world.

Yet, even with its undeniable potential, VR will still be limited by the act of leaving the real world to enter a virtual one. This is not something you do casually, in a similar way to how you can glance at your phone while waiting in the queue at a coffee shop. Once in VR, you can no longer see where you really are, and you start to forget where your physical body is. This leaves you vulnerable, which means that you need to start in a place where you are alone and safe. If you're exploring a virtual alien planet, the last thing you need is for a little sister or brother in the real world to run in and kick you where it hurts. VR also requires an open space of sufficient size for you not to keep bumping into furniture. Access to such empty space is becoming increasingly rare, especially as living in smaller and smaller city-centre flats becomes more common. The technology will get cheaper and the head-mounted display will get lighter and feel less like a scuba mask, but the act of mentally leaving the real world means that VR may never become a casual part of our daily lives. 3D televisions failed to take off with the public for a milder version of the same reason.

Augmented reality (AR), however, can be used more casually. In some ways AR is the opposite of VR. Instead of allowing you to enter a computer-generated world, it allows the digital realm to enter the real world.

Scott had talked excitedly about AR when I visited him. 'Augmented reality isn't a completely immersive new world like VR,' he explained. 'You stay in the real world, you can see the room around you, but where you are has been changed and improved. You still have an understanding of reality, you're still in that space, but you're augmenting it with things that shouldn't be there.

'At the moment AR glasses like Microsoft Hololens and Magic Leap aren't good. The field of view is too small. You can only see stuff appear in a box directly in front of you. But in two years' time? When Apple bring out their own glasses, then the augmented-reality experience is on. Apple aren't going to settle for a tiny field of view, I know that for a fact. They've obviously got a new way of doing immersive AR – of course they do, they're

Apple, they've got more money than anyone on the planet. They'll have a headset where things can happen all around. That's coming. But in another three years after that, John? I have absolutely no idea what it's going to be. Whatever's coming next, I can't see it yet. Nobody can. But it's coming.'

The best-known experience of AR is the smartphone game *Pokémon Go*, which became a worldwide fad in 2016. As with all Pokémon games, the aim is to find and collect a range of imaginary critters, which you then train and send out to fight other Pokémon. What was new about *Pokémon Go* was that the creatures were found in the real world, and to collect them it was necessary to leave your house. Your phone's camera would show the real streets and houses around you and the phone would superimpose the animated Pokémon on top of that view, if you were lucky enough to find one. It was not a sophisticated process, for the Pokémon could not interact with the world in any way, but it was effective enough for a generation of teenagers and younger kids to leave their bedrooms and voluntarily suggest going for a walk. What the potential of AR glasses offers is not seeing Pikachu on the pavement ahead when you hold up your smartphone, but simply seeing Pikachu on the pavement in front of you, behaving independently and interacting with the environment.

For this to happen will take an engineering advance in the design of AR glasses. As Scott has pointed out, glasses currently in development have a limited field of view. This is fine for seeing virtual objects appear on a desk or table that you are looking at, but it falls short of remaining believable when you move around the room. The belief that this problem will be overcome in the near future could be nothing more than an arrow-flight projection, of course, but it is worth exploring. If convincing affordable AR glasses do arrive on the shelves then their impact has the potential to be even greater than the arrival of smartphones.

These potential glasses would not have a built-in camera, so they would not be seen as a privacy risk in the same way that the much-criticised Google Glass prototype was. They would simply sit on your nose like regular sunglasses and, through wi-fi and the benefits of cloud computing, enhance and enchant the room you are in. If the people you are with are also wearing glasses, you could all experience the same shared hallucinations. The first versions of commercial AR glasses will no doubt be a little clunky, but it may not take too long before they become as stylish as sunglasses and branded by fashion companies such as Ray-Ban. Some futurologists insist they will not remain as glasses for long and will evolve into smart contact lenses.

If the technology does advance in this way, you could then share your home with immaterial objects and creatures which would appear consistent and three-dimensional, if perhaps a little misty and transparent, when you walked around them. The initial uses of the technology include games such as *Minecraft* or *Street Fighter* played on the kitchen table instead of a screen, or games where you run around your own home shooting at the virtual aliens or zombies breaking in. Demos such as these are fun, but they have been criticised as being little more than novelties. If AR is to catch on, it will need a 'killer app' to make it truly valuable. My hunch is that this will appear when the advances with AI virtual assistants are merged with the concept of pets.

With the admitted exception of my cat, pets are great. There is something valuable about sharing your home with something non-human. AR could offer virtual pets that are playful and entertaining, but which never leave a dead bird at the top of the stairs or crap behind the sofa. Your room could become a virtual aquarium, with immaterial fish swimming around the lights, while your children play with a cat-sized pony in the back garden. These AR critters would not be limited to real-life animals. You could share your home with established characters such as Smurfs, Porgs or Ant-Man. Buzz, Woody and the cast of the *Toy Story* films need no longer pretend to be inanimate toys when you enter the room but could continue to play in your presence. The Hogwarts ghosts from the *Harry Potter* books could float through the walls of your hallway.

Virtual pets like these would hide under the table and fall off the back of the sofa just like any puppy, but they would not chew the furniture or rack up large vet bills. With a haptic glove, you would be able to stroke these critters and they would not only look real, but they would feel real as well. But simple virtual critters like these are only the beginning of the potential that the merging of AR with AI offers.

Personal AI assistants are currently ethereal things. Alexa, Siri, Cortana and Google's nameless assistant will answer your questions, turn the heating down, play whatever music you request and arrange your social life by communicating with your friends' digital assistants. You can talk to them, but they are invisible, and talking to invisible things still feels a little unnatural. With AR, there will be no limits on the visual form these currently invisible characters might take. It's not hard to imagine a situation similar to Philip Pullman's *His Dark Material* books, in which everyone is accompanied by their own 'dæmon' that takes the shape of an animal of their own choosing. You can imagine how an anxious generation who avoid talking to shopkeepers would be quick to adopt such technology.

It's interesting to speculate how our relationships with AI assistants would change if they became visible digital spirits. What will their impact be on our culture, and what sort of careers and industries will they spawn? If these AR dæmons stop being servants and become companions we have relationships with, things could become interesting or troubling, depending on your point of view.

Who will be the first person to demand the right to marry their digital companion? Will sufferers from the schizophrenic family of illnesses be comfortable with, or distressed by, these unreal entities? Could such creations be legally compelled to share your secrets with the police, or your employer? It will take time and a lot of trial and error before we can answer questions like these.

In 2016, Disney used what were then state-of-the-art computer graphics to create photorealistic versions of young Carrie Fisher and Peter Cushing in the *Star Wars* movie *Rogue One*. For this breakthrough, they were nominated for the Best Visual Effects Oscar. But by the time the film was released on DVD, members of the public were using freely available apps that utilised a technique called Deepfake in order to create their own digital Carrie Fishers. Many people find the Deepfake version of Fisher more convincing than the professional one used in the film. Deepfake uses machine learning to generate a 3D model of a person's face from a collection of still photographs. It quickly became notorious when people used it to add the faces of famous actors to pornography. This was seen as demeaning to those famous actors and actresses, although it was probably more insulting to the porn stars who had their faces replaced by someone 'better'.

AI can already mimic people's voices to an impressive degree. Creating a 3D representation of someone's face has become relatively simple. Training an AI to mimic someone's personality would be more a question of smoke-and-mirrors trickery than a genuine re-creation of a mind, but it could be done well enough to pass in most situations. Given all this, a virtual character could look, sound and act like almost anyone.

In the same way that fans of movie stars or musicians might set their idol's image as the wallpaper on their phone, they would now be able to 'skin' their personal assistants as their own hero or heroine. For celebrities, selling virtual versions of themselves may be a lucrative sideline. AI could conjure up a digital spirit that looked, sounded like and to some extent acted like anyone at all, who would appear to be in your own home and whom you could converse with. All this has consequences.

We can be dismissive of such technology simply by saying that virtual entities aren't real and we shouldn't treat them as such. But if they become a larger and more integrated part of our lives, that distinction will seem less important. In the view of Edward Castronova, an American academic who specialises in virtual worlds and game design, this is already happening. 'The term "virtual" is losing its meaning,' he has written. 'Perhaps it never had meaning. The things happening online have always been literal human things [. . .] the allegedly "virtual" is blending so smoothly with the allegedly "real" as to make the distinction increasingly difficult to see. There's nothing revolutionary in this. It's merely a recognition that these things were always as real as anything else in the human culturesphere.'

We once thought that computers were things we would enter, but it may be that we had that the wrong way around. It is the digital spirits who will enter the real world, and enchant it, as the line between the physical and the digital becomes increasingly blurred. As we create this digital realm we will be, effectively, exploring our own imaginations. Our world could be anything we can dream up. For all the danger, there will be great joy to be had, as well as excitement, surprise and a lot of laughter. We are looking at a creative opportunity unsurpassed in human history.

Some people may find the idea that our future is not in outer space, and that we are confined to the Earth we already know, to be disappointing. But the Earth is no longer just the material world. The virtual world is already becoming as much our home as the physical. Because AI looks set to remain our tool rather than evolving into some all-powerful being, this is an arena that we will control. We are free to explore a dazzling world of unlimited creativity, if only we are able to trust the motives of those creating the digital world.

AR and VR, for all their potential, teach us something important about the future. These technologies serve to highlight what the virtual world is lacking. By filling our homes with the riches of the virtual imagination, we discover the extent to which trust and human relationships are what we truly value. When we imagine the future ahead of us, we need to remember that these are the things we will increasingly focus on.

7.
PSYCHIC POLLUTION

1.

When my children were in junior school they used to talk enthusiastically about their music lessons, but the stories they told sounded deeply odd. For example, they said that their music room contained an instrument so rare that there were only three of its kind in the world. Or they explained how their teacher, Dr Bramwell, had found an injured robin which he kept in his bag while he nursed it back to health. Nobody else was allowed to see the robin, because it was shy. Dr Bramwell also put a notice up around school saying that he'd found an ear which, judging by the size, he believed belonged to a Year 3 pupil, so if anyone had lost an ear they should see him. When my children sang songs he had taught them, they all seemed to be about animals getting run over.

David Bramwell – he is not a real doctor – has since left teaching and is now a writer, public speaker and the presenter of unusual documentaries on BBC Radio 3 and 4. He also produced the podcast series that accompanied my previous book, which gave me further insight into his unorthodox behaviour. For example, I now know that there were Yamaha keyboards in the music room, and it pleased David to spend years teaching children that Yamaha was pronounced 'Yer-*MAR*-har'. I doubt there are many people who would both think of doing this and find the idea so appealing that they would keep up the pretence for years. If you ever encounter someone convinced that the company is called Yer-*MAR*-Har, it's a safe bet who their childhood music teacher was.

The incident he seems most proud of was the time when, after growing a beard over a half-term holiday, he returned to work and told the children that he was his twin brother Steve, a dairy farmer from Cornwall, and that he would be taking lessons while David was on holiday. Not one child questioned why a dairy farmer was now teaching music, or who was looking after the cows. He spent the next fortnight taking lessons as normal, except that he spoke in a deeply unconvincing Cornish accent and pretended not to know any of the children's names.

I asked my son Isaac what he remembered of Dr Bramwell as a teacher. He replied, 'He was a bit crazy, but I liked him, he was nice.' My daughter Lia said, 'His room was in the basement, and he always had incense burning, so when I think of him now he's like a goblin. He would do stuff like show us that film about The Rutles but not explain that they were a spoof band. But after I left and found out that he was lying to us and that he's not a real doctor, I became a bit freaked. I mean, I can't trust anyone now. It's all lies.'

I meet David for a drink on a wet Sunday evening in January. It is not his unorthodox teaching that I want to talk to him about, but something related to the problem of trust in the digital world. As you may have sensed from his approach to education, David can seem a little removed from the normal everyday world. I know that he religiously avoids the news, and I wonder if this contributes to the sense of his otherness.

7. PSYCHIC POLLUTION

'I'd recommend avoiding the news to anyone,' he tells me. 'It improved the quality of my life. I am much happier.'

I ask him when he started.

'It was sometime around the beginning of the Millennium, I think. I used to listen to news on the radio in the morning and read newspapers, all the usual stuff. But I had a growing realisation that it wasn't benign, and it wasn't healthy. Keeping up with the news is supposed to tell you what the world is like and what's going on, but it really doesn't. We forget that the news is not how everyday life is experienced by most of us. It is the exception to the rule. Because we are bombarded with sensationalised daily horror stories there's a very real danger of developing a misanthropic view of humanity.

'Go into a supermarket or a newsagent and look at all the papers together. Virtually every headline will be about violence, disaster or a scandal. When these kinds of stories become our daily mantra, is it any wonder so many people think of the world as a terrible place? Tabloid headlines in particular are designed to trigger our sense of righteous indignation. Rather than encourage compassion, they offer an oversimplified view of the world with heroes and villains. The more shocking the headline the better. "If it bleeds it leads" is the old maxim in journalism. It's a proven formula for good sales. In 1991, it was discovered that the community of Damanhur in Italy had built an underground temple the size of St Paul's Cathedral in secret. I've seen it and it's the most astonishing place I've ever visited. It should have been the news story of the decade. Almost three decades on and it remains virtually unknown. Good news has a habit of not getting out.'

I can't argue with any of this. Having occasionally written articles for newspapers myself, I know the feeling of horror that comes when you discover what headline the editor has given your words. It will be a title chosen to get the biggest reaction, rather than to reflect what you're saying.

David continues. 'I didn't want to spend every day worried and angry, so I made the decision to stop consuming the news. Instead of putting on Radio 4 in the morning, I now put on a podcast or an audiobook, or I listen to music. On a really good morning, I'll spend that time playing an instrument. I use [the web browser plugin] Facebook Purity to hide the news-trending feeds that appear on Facebook, and I generally clear news off anything I can. I find out about what's going on from talking to people. If there's something important happening, then people discuss it with you. The healthiest way to engage with the news is to have conversations about it, though I realise this is letting other people do the dirty work for me.

'Avoiding things like the free newspapers on the Tube – that's a hard one. As an ex-smoker it's like resisting a cigarette. It's so tempting when I see them and look over someone's shoulder and see that Boris has been putting it about again. I have to resist getting pulled in. It's a discipline I have to keep working at. Sometimes the news is inescapable – it's on giant screens in so many of our train stations and public places. When I'm sat at my dentist's, waiting fifteen minutes to go in, I'm bombarded with bad news on a screen, while a banner with a different bad-news story scrolls along the bottom of the screen, so I can absorb two horrific stories in one. I've spent time in Vegas and Reno, and the first thing you notice is that slot machines are ubiquitous. They're in the banks, supermarkets, even the airports. Over here, it's headlines. There's something insidious about both. They're designed to be addictive.'

'Are you open about this?' I ask. 'Is it something you are proud of?'

'I'm very open about it. I've written about it in *The Idler*, and I've given talks, and generally the reaction has been very positive.

I'm always surprised that I'm not attacked more because of it. Even hard-line news junkies get it. I remember I once did a TEDx talk about this, and I was sat next to one of the sponsors of the event. And in the interval one of us brought up the talk. And he said, "Well, I suppose there's some benefit in what you're talking about but you're in a luxurious position where you might not have to engage with the news. But for some of us, it's our jobs, we've got no choice." And I looked at him and he was just so miserable. He had such a hangdog look about him. I just knew that twenty minutes on the bagpipes in the morning instead of the *Today* programme would have been so much better for his soul.

'But, that said, since I did my talk I've started to see articles by other writers popping up, saying much the same thing. There's usually one every few years, and I think they're getting more frequent. There's a larger pool of research now that backs it all up, so it seems more like science and less like one contrarian old bugger tilting at windmills. I think people get it quicker now. It doesn't come across as quite as strange as it used to.'

The research that David is referring to covers many different aspects of media consumption. A 2017 American Psychological Association survey, for example, found that 56 per cent of adults thought that following the news regularly caused them stress. A 2012 study by Fairleigh Dickinson University revealed that people who watched MSNBC or Fox News knew less about international events than people who did not watch any news at all. Research by the University of Sussex shows that negative news broadcasts exacerbate viewers' own worries, making viewers unhappier and more anxious.

'What about voting?' I ask. 'After all, the whole idea of democracy requires citizens to have a reasonable level of understanding about what's going on.' David swigs his beer and nods.

'The questions I am always asked about it are around

irresponsibility, and whether or not I have the right to vote. That's the thing that keeps coming up. It's a tricky one. When it comes to elections I'll read manifestos, and I will tune in because I feel there's a responsibility there to be as well informed as I can in terms of who to vote for. After that I switch off again. But, then, look at what happened with Brexit. The more people consumed the news, the more misinformed they were. I definitely feel much less of a need to apologise for this after the Brexit referendum.

'I don't think it's irresponsible. You have to ask yourself: how does my knowing about these depressing stories help? Am I making the world better in any way or just getting angry and banging on about it with my friends? I have friends who obsess over Trump's latest tweets and share silly little memes and videos where someone is pooing on Trump's head. I'm not sure what it achieves, but it's a great way to give yourself an ulcer.

'Remember when Madeleine McCann was abducted? That story ran for months in the press. It was tragic, and the fear of child abduction is still in parents' minds, but that fear is not a true picture of the reality of child abduction. My friend Steve Colgan, who's an author, ex-policeman and former researcher for *QI*, was part of a team who looked at the statistical likelihood of abduction in present-day Britain. Using data from the last fifty years he and his team calculated that if you left your child outside your home unattended, statistically it'd take an average of ten thousand years before they were abducted. That's the reality of the situation, but following the news makes you think that things are far worse.

'These days there's a real interest in how we can improve our well-being, and there are whole industries springing up around mindfulness, and there's a lot of research being done in these areas. I've just read a book on the science of sleep, and there was a lot in there about the importance of finding ways to calm the mind, and not being in fight-or-flight mode all the time. My adrenalin levels were once fit to bursting from overdosing on news. The best thing I did for my mental health was to give it up. News really should come with a health warning: side effects can include anger, depression, frustration, righteous indignation and an unhealthy obsession with socialites, celebrities and royals. Yes, I'm relatively ignorant of what's going on, but my God I feel happier, and I couldn't go back. I wouldn't want to go back.'

2.

During the hundreds of thousands of years of *Homo sapiens* evolution, fat and sugar were relatively scarce. Getting a bit of fat and sugar down our necks when the opportunity arose was a positive survival skill, and evolution naturally favoured those with a fondness for such goodies. Countless generations later, we in the modern world are often unable to resist fat and sugar, even though they are now anything but scarce. The result of this is a huge amount of obesity, diabetes, heart disease and other health issues.

Likewise, after the cognitive revolution and the arrival of early language, there was not a lot of myth or narrative around. We had to make do with the stories of our tribe along with gossip about the relationships and behaviours of our fellow tribe folk. Because social integration had an evolutionary advantage, members of the tribe who were more involved with these tribal dynamics were more likely to pass on their genes. Tens of thousands of years later, we have a deep need for gossip, stories and all that makes up our circumambient mythos. We spend our evenings in front of screens, bingeing on the latest box sets from Netflix or snacking on the latest bite-size chunks of gossip from Twitter. We might think that watching current-affairs programmes like *Question Time* is worthier than watching soap operas such as *Coronation Street*, but the brain's reward system is unconcerned with this distinction, and they both satisfy the same cravings.

In the twenty-first century it is attention that is scarce, not narrative. Each individual little nugget of news is in danger of being overlooked because of the sea of alternatives. When attention is rare, it is necessary to become an attention seeker. News now shouts louder than before. It promises novelty and horror, and it knows that the easiest way to attract us is to appeal to our lower natures. When everything is louder than everything else, and every piece of news is announced as the most terrible thing ever, it can feel like the mythology we live in is undergoing something of a nervous breakdown. As the psychologist Steven Pinker describes the situation, 'every day the news is filled with stories about war, terrorism, crime, pollution, inequality, drug abuse, and oppression. And it's not just the headlines we're talking about; it's the op-eds and long-form stories as well. Magazine covers warn us of coming anarchies, plagues, epidemics and so many "crises" (farm, health, retirement, welfare, energy, deficit) that copywriters have had to escalate to the redundant "serious crisis."'

Our desire for hits of information became apparent when smartphones arrived. People now openly joke about how addicted they are to their smartphones, and how much time they lose playing with them. For many, the thought of losing their smartphone is far more distressing than the thought of losing their house keys or their wallet. It is telling that there is a word for the fear of losing your phone – nomophobia – but there is no word for the fear of losing your wallet. It is almost as if our smartphones have become part of the essence of us, which is why the thought of being separated from them is physically distressing. We treat smartphones as if they were an aspect of our souls.

Part of the reason for this is the neurotransmitter dopamine. This is a very useful brain chemical. It is involved in many complicated functions, which range from physical movement to behaviour reinforcement. It is this last role that gets the most attention, because when your brain anticipates that it is about to experience something pleasurable, dopamine makes you more likely to repeat that behaviour in the future. As a result, dopamine plays a role in establishing addictive, destructive behaviour. Should you find yourself endlessly scrolling through social media or clicking on links searching for new information which will give you a hit of pleasure, it is dopamine that drives you on.

Silicon Valley knows all about how this biological mechanism works. A company called Dopamine Labs, for example, promises to use an understanding of neuroscience to keep people returning to the apps on their phone. Their website tells you to 'Connect your app to our Persuasive AI and lift your engagement and revenue up to 30 per cent by giving your users our perfect bursts of Dopamine.' The company's founder, Ramsay Brown, studied neuroscience before building his company and is interested in what he calls 'brain hacking'. He told CBS's *60 Minutes* programme that 'a computer programmer who now understands how the brain works knows how to write code that will get the brain to do certain things'. As an example of how this works, Brown claimed that Instagram uses an algorithm which might hold back from telling you the correct number of likes your picture has received. Should you return to the app to see the reaction to your photo and feel disappointed, you'll then return later and be overjoyed to see the true numbers. Manipulating a user to feel first unhappy and then happy like this keeps them returning to the app, which means that they will see more ads. Apps like Facebook are 'engineered to become addictive', Brown said. 'It is the case that [there's an addiction algorithm]. Since we've figured out to some extent how these pieces of the brain that handle addiction are working, people have figured out how to juice them further and how to bake that information into apps.'

One telling detail about the software used by Dopamine Labs is the name they gave to the AI that manages all their data: Skinner. The software is designed to offer occasional rewards exactly like a Skinner box. Users are compelled to keep interacting with the apps, but think that they are doing so out of their own free will.

Withholding rewards in this way is the same principle used by slot machines, which are mathematically designed to provide the right amount of occasional minimal rewards to keep the player playing. In a location such as a Vegas casino where there are no clocks, windows or other indicators of passing time, the mathematics behind sporadic slot-machine payouts can keep gamblers playing for many hours at a time. In a similar way, social-media platforms also hook their users by only occasionally providing them with a nugget of information which excites them.

The same principle is used a lot in videogames and, in particular, in RPGs (role-playing games). In RPGs, the player's statistics and abilities are increased slowly but surely during lengthy play. So effective are RPG-style rewards in keeping gamers playing that RPG elements are now routinely introduced into non-RPG games. In the original *Tomb Raider* games, for example, the sense of exploration and discovery was sufficient reason for you to invest your time in the game. In modern *Tomb Raider* titles, extended play earns you points which you then allocate to weapons, fighting or survival skills. Many gamers are unhappy about the practice of adding RPG elements to everything, believing that such tactics are a cheap, easy and mindless way to hook gamers. It is much better, the argument goes, if games interest the player with surprise, mystery, story and other less cynical methods of creating an absorbing game. But doing that is difficult. RPG elements are both easy and known to work.

The open use of the concepts behind Skinner boxes would horrify someone like Jaron Lanier. But while Lanier and the other computer pioneers who wrestled with these ethical issues were members of Generation X, the Millennial generation that flocked to Silicon Valley afterwards were less interested in moral concerns. As the 1990s became the twenty-first century, Silicon Valley became a place where idealised dreams of building a better world were replaced by the ultimate goal of making money. As the former Facebook engineer Jeff Hammerbacher famously complained, para-phrasing Allen Ginsberg's *Howl*, the 'best minds of my generation are thinking about how to make people click ads'. In this culture, using the techniques of Skinner to control the users of apps, and making them spend far more time scrolling and clicking than they would otherwise choose, went from being self-evidently wrong to being an admirable business practice. In this atmosphere, a company like Dopamine Labs could not only call their AI Skinner, but openly announce that that's what they have done.

None of this was inevitable. In an alternative universe, there is a version of Twitter which has none of the numbers that tell you how many likes or retweets your comment has received. Twitter could have been a non-profit venture, like Wikipedia, and existed to provide a simple way to enable communication for everyone, whenever it naturally arose. But, in our universe, Twitter, Inc. listed on the New York Stock Exchange in 2013 and thus has to provide shareholders with growth in both user numbers and advertising income. How Twitter works has been changed numerous times since then to enable this. Algorithms now select what tweets you see or don't see, in an effort to generate reactions and keep you returning to the service. As long-term users of the service will tell you, the atmosphere on Twitter changed radically over this period. It originally felt like a global dinner party, where everyone was welcome and allowed to contribute. It now feels more like a rave in an asylum, where everyone is shouting about how it's everyone else who is crazy.

Most of us are in denial about the amount of time we waste on our phones. It always seems to be our friends and family who are the addicts, rudely staring at their phones in company, while we have our use more or less under control. To help shatter this illusion, there are apps you can download to measure how often you unlock your phone and how long you spend using it. I tried one and was so shocked at the results that I suspected it wasn't working properly. I downloaded another app in order to check, but this one's damning accusations were exactly the same. The irony of needing an app on my phone to understand how reliant I am on my phone was not lost on me.

3.

The art of making computers shape human behaviour has a number of names, including 'persuasive tech', 'behavioural design' or 'captology'. Stanford University has had a persuasive-tech lab since the 1990s. The front page of its website has the heading 'MACHINES DESIGNED TO CHANGE HUMANS'. It tells us that 'Yes, this can be a scary topic: machines designed to influence human beliefs and behaviors. But there's good news. We believe that much like human persuaders, persuasive technologies can bring about positive changes in many domains, including health, business, safety, and education. We also believe that new advances in technology can help promote world peace in 30 years.' I think this wild claim was meant to be reassuring.

The first major social-media insider to confess that our growing addiction to digital devices was not an inevitable side effect of the technology, but a deliberate goal pursued by the designers of software, was Sean Parker, the billionaire coder who co-created Napster and became the first president of Facebook. As he said in October 2016, 'The thought process that went into building these applications, Facebook being the first of them, was all about: "How do we consume as much of your time and conscious attention as possible?" That means that we need to sort of give you a little dopamine hit every once in a while, because someone liked or commented on a photo or a post or whatever. And that's going to get you to contribute more content and that's going to get you [...] more likes and comments,' he said. 'It's a social-validation feedback loop [...] exactly the kind of thing that a hacker like myself would come up with, because you're exploiting a vulnerability in human psychology. The inventors, creators – me, Mark [Zuckerberg], Kevin Systrom on Instagram, all of these people – understood this consciously. And we did it anyway.' He also remarked, 'God only knows what it's doing to our children's brains.'

Parker's public comments were a big deal. There was no culture in Silicon Valley of criticising the industry or its actions. As Jaron Lanier said in a 2018 interview with Vox podcast, 'not too many years ago I was still being criticised quite intensely for bringing up any point of difficulty in the online world [...] Oh my gosh I lost friends over that, not that long ago. Imagining how things can go wrong is an absolutely essential ability, and it was just forbidden in Silicon Valley culture.' But Parker's confession broke the spell. It led to many other Silicon Valley insiders speaking out.

Facebook's former vice president for user growth Chamath Palihapitiya told a conference at Stanford Business School in December 2017 that he felt 'tremendous guilt' that he helped create 'tools that are ripping apart the social fabric of how society works'. He explained that 'the short-term, dopamine-driven feedback loops that we have created are destroying how society works'. Roger McNamee, venture capitalist and mentor to Mark Zuckerberg, wrote articles in the *Guardian* and *USA Today* highlighting the need for Facebook to address the problems of addiction. 'I am a tech investor, and Facebook is by far my largest investment,' he began. 'Still, for the past 15 months I have been pushing Facebook to sacrifice near term profits. The reason? I want them to address the harm the platform has caused through addiction and exploitation by bad actors. Government watchdogs barely regulate the technology sector in the United States, so investors like myself have a big role to play.'

Concerns such as these led to the founding of a campaigning organisation called the Center for Humane Technology, which aims to help us move away 'from technology that extracts attention and erodes society, towards technology that protects our minds and replenishes society'. One of the organisation's founders is the former Google designer Tristan Harris, who highlights the addictive nature of certain social-media functions, such as Snapchat's streaks feature, which records how many days in a row two friends send each other picture messages. 'It creates this false sense of, "I've got to keep this thing going",' Harris said. 'It starts to really go deep into them and actually redefine their identity and their meaning of friendship. They think that they're not friends if they don't keep that streak going.' In the world of drama that is teenage dating, the implications of breaking a Snapchat streak can run deep indeed.

The problems of technology addiction were not news to the great and the good of Silicon Valley. Steve Jobs famously didn't let his children use an iPad, telling a *New York Times* reporter that 'We limit how much technology our kids use at home.' His successor at Apple, Tim Cook, told the *Guardian* that 'I don't have a kid, but I have a nephew that I put some boundaries on. There are some things that I won't allow. I don't want them on a social network.' Chamath Palihapitiya took a similar attitude, saying that 'I can control my decision, which is that I don't use that shit. I can control my kids' decisions, which is that they're not allowed to use that shit.' All this is reminiscent of the attitude of an executive from the cigarette company R. J. Reynolds, who when asked why none of his company's senior executives smoked their own product replied, 'We don't smoke the shit, we just sell it. We reserve the right to smoke for the young, the poor, the black and the stupid.'

None of this is truly new. Mass-market advertising is a blunt tool compared to the individually tailored experience of social media, but the history of advertising is still a history of psychologically manipulating the wider population to do certain things, and to believe that they are doing those things out of their own free will. Advertising, of course, has numerous benefits for the economy. It informs consumers about products that are available, and it helps increase trade and commerce. When proceeds from advertising funded newspapers, television shows and magazines, it used to help create culture, although the rise of the internet has caused this funding to fall dramatically. Some believe that the creativity on display in advertising was a form of culture itself. The problem is, however, that advertising also has a dark side. It utilises its understanding of how the brain works to make huge numbers of people spend money on products they did not need and would not have bought otherwise, and it ignores the negative effects that it has on the psyche of those it manipulates.

An example of the techniques it uses is the 'pack shot' at the end of an advert, the close-up image of the product being sold, which is usually preceded by an image of an attractive, confident, happy person. By cutting from this positive human image to the product, a connection is made in the mind of the viewer which creates an association between the product and those positive emotions. This technique is far from new and was pioneered by the Russian filmmaker Sergei Eisenstein in the early part of the twentieth century. Eisenstein is the father of film montage, and his techniques have been used extensively by propaganda filmmakers ever since.

Tricks like this are not as harmless as they may first appear, because the associations which are built in the mind are bi-directional. Advertisers like images of children hugging their parents, a first kiss, beautiful people, romance, luxury, pets or happy families, because their audience have positive associations with their own versions of such memories. They are, after all, the very things that make life precious. By cutting from such personal moments to an image of the product, they allow that product to take on some of the positive qualities associated with the viewer's children, family or romantic life. But because those connections are bi-directional, the viewer's personal memories also become associated with what is being sold. If that product is cheap junk, then the positive aspects of the viewer's life become associated with cheap junk. As a result, the viewer's internal emotional life becomes poorer. The effect is only slight, of course, but, when we are exposed to as many adverts as we are today, it builds up.

A particularly troubling example was an advert McDonald's aired in the UK in May 2017. It featured a boy trying to come to terms with the death of his father. The viewer's empathy for this child was then manipulated by the advertisers to make them think positively of McDonald's Filet-O-Fish burgers. The advert resulted in bereavement charities receiving huge numbers of calls from parents of bereaved children, who had been upset by the advert. For young children of deceased parents, their memories of their late father or mother are precious because they are so limited and can never be increased. To contaminate them with something so crass, therefore, is extremely cruel.

McDonald's withdrew the ad after they received widespread criticism for exploiting bereaved children. 'It was never our intention to cause any upset,' they said in a statement. 'We will also review our creative process to ensure this situation never occurs again.' The implication here is that the professional advertising people involved did not understand their craft and did not know what they are doing. Personally, I suspect that this is not the case. It seems more likely to me that the advertisers themselves were not bereaved as children, and that they designed the advert to manipulate the emotions of an audience who were also predominantly not bereaved as children, but who would feel sympathy for a suffering child. Certainly, the associations the ad creates in those who have lost a parent as a child are very different to those needed to create the desired increase in fast-food sales. Speaking as someone who lost his father at the age of three, I now think of a McDonald's Filet-O-Fish burger as the single most dismal and depressing item of food available.

4.

That advertisers' psychological manipulation can cause harm has long been understood, but there have not been any significant campaigns to prevent the psychological pollution it creates. There's a considerable amount of research which shows that activities that use screens, be that for going online, using social media, texting or watching TV, make people noticeably less happy than non-screen activities such as meeting friends, playing sports, reading or even doing homework. Teens who spend more than five hours a day online, for example, are twice as likely to be unhappy as those who spend less than an hour a day at screens. Videogames, incidentally, are not as harmful as social media.

Because of this I decided to delete Twitter and Facebook from my phone, and to also switch off the notifications that ping whenever an email is received or whenever an app feels it is not getting enough attention. The red dots and urgent noises that constantly harangue everyone with a smartphone were invented by behavioural designers to compel people to act, so they had to go.

Afterwards, I became less likely to take out my phone whenever I sat down, to mindlessly flick through the noise in search of a dopamine hit. My phone use dropped dramatically, from around two and a half hours a day to just under an hour. I even started occasionally leaving the house without my phone, something that would have been unthinkable before. I no longer had the nagging background anxiety about the state of my phone's battery, or the amount of data I used. I still used social media on my laptop, so I didn't feel like I'd cut myself adrift from the world. Ultimately, it is a matter of limiting rather than ending screen use. Rather like alcohol, it has been suggested that a small amount of use seems to be better for you than total abstinence. A similar approach has been taken by the man who created Facebook's 'like' button, Justin Rosenstein. 'I no longer have news apps on my phone or pretty much anything that could distract me,' he has said. 'If you're trying to lose weight, don't keep cookies in your pocket.'

There is a negative feedback loop involved in social-media use. The more you use it, the more you think that you're missing out on what's going on. Cutting back on these products worked well for me, and the overall result of deleting those apps from my phone has been to leave me noticeably more content. Like David with the news, I have no desire to go back.

I found limiting myself like this to be simple and painless, but then I do not seem to have an addictive personality. I am completely uninterested in gambling, for example. Back in the 1990s, before my children were born, I used to smoke cigarettes. I enjoyed this, but I worried that it had become such an automatic habit because I often lit a cigarette and smoked it without even noticing I was doing so. If nothing else, this was a waste of money. In response, I decided I would only smoke during the winter months. Each year I quit smoking on the spring equinox and took up the habit again at the autumn equinox, after which I found that I appreciated the cigarettes far more than before.

For the few years I followed this system it achieved its aims brilliantly. I got far more enjoyment from smoking, and I did not feel like I was taking it for granted or wasting my money. I did not find sticking to this routine difficult. The only problem was the reaction of other people, which was universally negative. Smokers did not like the casual way I could stop smoking every summer without relapsing. In the eyes of some smokers, this seemed to be showing off. Non-smokers, meanwhile, became extremely angry that I was planning to start smoking again when winter came around. Anyone planning on starting smoking, clearly, was deranged. Given these negative reactions, it was almost a relief when I became a father and stopped smoking for good.

Researchers who are interested in the neurological basis of addiction pay a lot of attention to an area of the brain called the nucleus accumbens. This plays a significant role in the processing of pleasure, reward, motivation, reinforcement and other functions that are integral to addiction. There are two main types of dopamine receptors in this part of the brain. These are, somewhat unimaginatively, called the D1 dopamine receptors and the D2 dopamine receptors.

Dopamine is often portrayed as the brain's reward chemical, but this is not technically correct. When our brains provide us with the experience of pleasure, it is usually as a result of naturally occurring opioids, such as endorphins, binding to one of the brain's opioid receptors. Dopamine is not the chemical that produces that hit of pleasure, but it is the chemical that reinforces the behaviour which caused that opioid fix in the first place. It makes you more likely to repeat that behaviour in the future.

There is a difference between the two different families of dopamine receptors. D1 receptors are stimulated by things like food and material items. They are also hijacked by drugs like cocaine. Stimulation of D1 receptors, in other words, reinforces material desire and makes us want stuff. D2 receptors, in contrast, are activated by social interactions, and their stimulation makes us increasingly likely to exhibit social behaviour. It does not make us want stuff, it makes us want people.

Those who have fewer D2 receptors than average are more prone to substance addiction. It has been suggested that this is because they attempt to compensate for the lack of D2 stimulation by overstimulating their D1 receptors. Given my casual attitude to gambling, smoking and smartphone social media, it seems likely that I am lucky and have a healthy crop of D2 receptors in my nucleus accumbens.

Recreational drug use makes the problem of a relative lack of D2 receptors worse, because chronic substance abuse and excessive D1 stimulation actually reduce the number of D2 receptors. As a result, drug users can find it increasingly difficult to find satisfaction in social encounters. This is part of the reason why heavy users increasingly become isolated from the rest of society and are more prone to suffer from the mental-health issues linked to social withdrawal.

It is tempting to intuit a role for D1 and D2 receptors in the subjective experience of using Twitter over the years. When the service started, its design seemed to increase social behaviour. Something about using Twitter in the early days made you want to talk to strangers and make friends, as if your D2 receptors were being stimulated. This has gradually changed as the service has been altered over the years. Now people you don't know on

Twitter seem to be broadcasting rather than talking, and the desire to make friends with strangers is considerably less. It is as if the service has been re-engineered since it floated on the stock market, changing it from something that stimulated your D2 receptors into something that is more focused on stimulating your D1s.

To find examples of social media engineered to stimulate social D2 receptors over D1s, you need to look away from Twitter, Facebook or services used by older people, and instead look at social media preferred by the young. Snapchat streaks, as we mentioned earlier, are designed to encourage repeated daily use of the platform, but the addictive nature of streaks seems to be created more by stimulating the social D2 receptors, not the hungry D1s.

This is not, admittedly, an easy subject to produce clinical research about. Scientists interested in the role of dopamine in research frequently use animals such as rats. They feed them with chemicals, observe the resulting behaviour and then dissect the rat to examine the brain. Getting rats addicted to speed or cocaine is easy enough, but you're not going to get rats hooked on Twitter or Instagram. Discussing the role of D1 and D2 receptors in social-media use is therefore something of an inference, but it is probably not too controversial to suggest that advertising-led consumer culture has historically learnt how to stimulate D1 receptors and paid less attention to techniques that stimulate social D2s. The rise of isolating individualism in the twentieth century certainly mirrors the rise of marketing. Likewise, a younger generation raised on networks and exhibiting greater social empathy does seem a better fit for social-media features like Snapchat streaks, which encourages more social, D2 receptor-like behaviour.

The hormone oxytocin also has a role to play in social interaction and behaviour. Oxytocin is often called the love drug or the cuddle chemical. It is linked to empathy, trust and generosity. It bonds couples after orgasm and is released in great quantities in mothers after childbirth, in order to create a strong maternal bond. Oxytocin has long been viewed as a great thing, and you would expect to find a lot of it released in the minds of a generation as empathetic and socially linked as Generation Z. But we are only starting to realise that it may have a dark side.

There is increasing evidence that the chemical plays a role in negative social interactions as well as positive ones, and that it can amplify bad experiences just as much as good ones. In a 2017 study, researchers at the University of California subjected mice to the mouse equivalent of bullying and found that this activated oxytocin receptors in an area of the brain associated with social anxiety. This resulted in oxytocin-driven social isolation, as the poor, picked-on mice began avoiding their peers. This echoed 2013 research by the private company Northwestern Medicine which found that, during and after the experience of social defeat or trauma, oxytocin targeted a specific area of the brain that reinforced fear-based memories.

Research such as this makes it tempting to speculate about a role for oxytocin in the increase in social anxiety shown by post-Millennials. While their displays of empathy and emotional intelligence suggest an increase in socially triggered oxytocin, it may be that the same oxytocin means that negative social experiences have a far stronger impact on their mental health.

There is also an increasing body of evidence that links oxytocin to tribal behaviour, to the extent of generating hostility towards outsiders. Chimpanzees, for example, receive a rush of oxytocin before going into battle with rival groups. In a Dutch study, students playing a hypothetical life-or-death game became more likely to choose people with Dutch names to save, and Arab names to kill, once they had been given an oxytocin nasal spray. While we tend to think oxytocin makes us more loving, it seems that this only extends to those we identify as being part of our tribe.

Outsiders are seen as a threat to our beloved tribe, and must be dealt with accordingly.

Neurotransmitters like dopamine are only part of the story of smartphone addiction. In 2017, a team at Korea University in Seoul studied the brains of teenagers whose use of technology was detrimental to their lives and compared them with the brains of teenagers who reported no problems with tech use. They found that those with problems had a greater amount of the neurotransmitter GABA and less of a chemical called glutamate. GABA helps regulate anxiety and slows down brain signals, while glutamate makes neurons more electrically excitable, so it appeared that excessive smartphone use had a calming, soporific effect on the teenagers. After the teenagers underwent a course of cognitive behavioural therapy to control their smartphone use, the amount of these chemicals returned to normal levels. It was as if they were using their technology to self-medicate the anxiety caused by the oxytocin-rich social lives that their technology had enabled. If this is the case, then here is another feedback loop keeping us glued to our phones.

5.

Facebook has had a trust problem from the very start. Back in 2004, when its founder Mark Zuckerberg was 19, he referred to the first few thousand members of his fledgling Facebook as 'dumb fucks' for trusting him.

Ten years later, a paper by Cornell University researchers showed that the teenage Zuckerberg had a point. The paper, called 'Experimental evidence of massive-scale emotional contagion through social networks', detailed an experiment conducted on 689,000 Facebook users without their knowledge or permission.

This experiment manipulated their Facebook feeds to show either more negative content or more positive content. By analysing the later Facebook updates of these people, the researchers showed that the tone of the social media people were exposed to had influenced their mood. Exposure to negative content, in other words, made people more unhappy, depressed or angry. They were more likely to go on and create further negative content.

Control of the algorithms that decide what each Facebook user can see gives Facebook extraordinary power over billions of people. It is easy to imagine how this could be used to influence an election by making voters more angry or fearful on the morning of election day. We will never know the full extent of the social impact of making Facebook's algorithm more negative, but when hundreds of thousands of people have been nudged into a bad mood it seems plausible that some of their friends, family or co-workers bore the brunt of this.

In 2017, *The Australian* newspaper published an internal memo written by two senior Australian Facebook executives. It claimed that Facebook's algorithms could identify and exploit Australians as young as 14 and allow advertisers to target them at their most vulnerable, including when they feel 'worthless' and 'insecure'. Facebook, it seemed, was not only monitoring the mental states of children, but selling that information to advertisers so that they could deliver an advert perfectly targeted to appeal to those vulnerable people. Facebook not only had the power to make people unhappy, it also had the ability to profit from that unhappiness.

Since 2016, Facebook has become synonymous with politically manufactured fake news. This is a different form of psychological manipulation, designed to control your opinions and voting intentions rather than your consumer choices. In the three months leading up to the American presidential election of 2016, fake news stories were more widely shared on Facebook than real ones. As a

Buzzfeed investigation revealed, 'During these critical months of the campaign, 20 top-performing false election stories from hoax sites and hyperpartisan blogs generated 8,711,000 shares, reactions, and comments on Facebook. Within the same time period, the 20 best-performing election stories from 19 major news websites generated a total of 7,367,000 shares, reactions, and comments on Facebook.'

The reason these stories performed better was that they were designed to do so. They were invented with knowledge of our psychological biases and created to appeal to us emotionally rather than intellectually. They were an appealing weapon for political campaigns that wished to be associated with a positive message, because their origins were hidden. When Channel 4 News conducted undercover filming of executives from Cambridge Analytica (CA), the notorious political-data analysis company which had access to Facebook data, it recorded the managing director of CA Political Global, Mark Turnbull, explaining that 'We just put information into the bloodstream to the internet and then watch it grow, give it a little push every now and again over time to watch it take shape. And so this stuff infiltrates the online community and expands but with no branding – so it's unattributable, untrackable.' We don't know where widely shared fake stories, such as those linking Democrat politicians to a paedophile ring in a pizza restaurant, originated from. But we do know that someone, somewhere, found it advantageous to make things like that up, knowing how Facebook works.

All this has led to a great deal of criticism of Facebook and the wider Silicon Valley culture that it represents. The tech world has very quickly gone from being perceived as being made up of brilliant, innovative pioneers to tax-dodging, competition-crushing, anti-democratic privacy-invading monopolies who are not to be trusted. Speaking at the Davos World Economic Forum in 2018,

George Soros, one of the world's wealthiest investors, described how 'Facebook and Google have grown into ever more powerful monopolies, they have become obstacles to innovation, and they have caused a variety of problems of which we are only now beginning to become aware [...] Social media companies deceive their users by manipulating their attention and directing it towards their own commercial purposes. They deliberately engineer addiction to the services they provide. This can be very harmful, particularly for adolescents. There is a similarity between internet platforms and gambling companies.' He went on to warn that 'The internet monopolies have neither the will nor the inclination to protect society against the consequences of their actions. That turns them into a menace and it falls to the regulatory authorities to protect society against them.'

Companies like Facebook are aware that they now face an existential threat. They will remember the fate of the original AT&T, which was forcibly broken up by the US government into a number of smaller companies, following an anti-trust suit in 1984. Breaking up companies that have become too big is a blunt tool, but there is a strong feeling that things can't go on as they have.

Facebook has responded by trying to limit its more problematic behaviour in an attempt to demonstrate that it can be trusted to self-regulate. It has reworked its algorithm to present the user with less fake news and ads and more social interactions with their friends, and it has been running TV ads to make people aware of this change. It no longer wishes to be thought of as a place where people are whipped up into a mob by anti-immigration 'Britain First' memes. Instead, it wants to be thought of as the place where you can see your cousin's new puppy.

It may not be the threat of prosecution that has caused this change. It may be something which, for Facebook, is far more serious. While the number of older users continues to grow, teenagers have been abandoning the platform in massive numbers. In 2018, just 51 per cent of Americans aged 13 to 17 used Facebook. This is a dramatic plunge from the 71 per cent who used the social network in 2015. This sort of collapse is an existential threat to the company's long-term future.

The way the platform can be used to trigger negative emotions works as a powerful addictive drug on the Baby Boomer generation, yet this repels the young. This is an example of how the generational change we are witnessing can have huge effects on the wider culture, as it is already causing changes in a multibillion-dollar corporation. Mark Zuckerberg was always open about how the changes they were making would affect their profitability, but investors didn't really seem to believe him until an earnings report and conference call with analysts in July 2018. Suddenly investors realised that he was serious, and $119bn was wiped off the value of the company in a single day. This was the most dramatic share price decline of any company in US stock market history.

This gives some indication of how important Generation Z is for the long-term future of the company. Facebook seems to have realised that the very things which made it one of the most successful companies in the world are exactly the reasons why post-Millennials want nothing to do with it.

6.

Given the biological aspect of addictive behaviour, the idea that people can just control and limit their use is clearly not going to be the way forward for everyone. How willing Silicon Valley in general will be to voluntarily reduce the addictive nature of its products, knowing that this will lead to smaller profits, is yet to be seen, but there is obviously a considerable degree of scepticism about how far it will go.

It is easy to assume that harmful activities such as these are destined to continue unopposed indefinitely. Individual adverts and clickbait news stories can easily pass as innocent, or even beneficial. The damage caused by observing a single, harmless-seeming advert is far from obvious, while the advertising industry's role in promoting economic growth is clear.

This is what makes the talk, from organisations like the Center for Humane Technology or from financiers like George Soros, about regulating the actions of companies like Facebook so significant. It may be that circumstances have suddenly become favourable for a reduction in this psychological manipulation. If we look back at history, we see that the need to regulate often only became apparent after a rapid increase in the scale of the problem. The move from passive, mass advertising to personalised targeted messaging may have created such a tipping point.

An example of the impact of scale is the ugly history of slavery. Throughout history, slavery was known in almost all parts of the world and was practised by many different cultures. Because slavery had always existed it was easy to assume it as part of the natural order of things, rather than something self-evidently morally wrong that needed to end. When Britain passed the Abolition of the Slave Trade Act in 1807, and the Slavery Abolition Act of 1833, this was historically unprecedented. But these acts only happened after the British Empire and others had industrialised slavery through the North Atlantic slave trade, causing the practice to expand massively. Only after the scale of slavery had exploded did the harm caused by the practice become so apparent that it could no longer be ignored.

Another example, at least in continental Europe, was war. There had always been war between European powers, and the idea that this could stop must have seemed no more plausible than wishing for an end to thunderstorms or droughts. But once war had become industrialised in the horror of the First World War, the idea that this had to be the 'war to end all wars' became openly discussed. The amount of death and suffering that tanks, chemical weapons and aerial warfare inflicted was far in advance of the earlier armed cavalries, and this ended the idea that war was something we must accept. The impact of this was not immediate, because resentment about the enforced settlement after the First World War eventually led to the Second World War. But when peace finally settled, international institutions were built that have made the idea of one European country invading its neighbour change from something that used to be a common occurrence to something now largely unthinkable. Once again, a long-established practice only became seen as unacceptable when it was industrialised and massively expanded.

This may be a similar situation that we're at now with psychological pollution. A few years ago, the idea that it could be controlled and regulated was almost entirely absent in our culture. Now, it has a strange sense of inevitability. If the reactions to deliberate psychological manipulation by social-media apps like Facebook do lead to less damaging practices, this will serve as a useful reminder that we can, as a culture, change course, and that there are occasions when we start to behave better. This is useful to remember, because most dystopian futures assume that we keep on doing all the things that are bad for us.

Some problematic behaviour is straightforward to change. When it was discovered that chemicals called CFCs in aerosols and refrigerators were accumulating in the upper atmosphere and causing a hole to appear in the ozone layer over the Antarctic, international governments came together and produced the Montreal Protocol. This was signed in 1987. The legislation phased out production of

CFCs and similar chemicals, and as a result the hole in the ozone layer has slowly started to shrink. It will still take another 50 years or so before the hole is completely healed, and the detection in 2018 of a mysterious source of CFCs from somewhere in China is a reminder that we need to remain vigilant, but international action does seem to be largely working as planned. Once the problem was identified, the solution was relatively painless because there were uncontroversial alternatives to CFCs use, and the change would not cause significant problems for the economy. Unfortunately, not all problems are so straightforward.

Climate-change legislation is considerably more difficult. The scientific consensus about what we must do to keep our climate within its historic rhythms is both expensive and politically difficult. In circumstances like these, you may assume that it will take a massive escalation in extreme weather and climate chaos in order for us to change course, in a similar way to how slavery and European war became unthinkable after they were industrialised. The problem is that by the time we get to that point, it will be too late. The delay between carbon entering the atmosphere and the climate changing is too long for our normal reactions to kick in. Even if we completely decarbonated our economies today, the climate would continue to change for decades to come. One of the problems with climate change is that we are just not good at reacting to something that occurs so slowly.

What role, then, will a changing natural world play in our future, and how will we react to it?

8.
FIXING THINGS

1.

As I walk over to Melinda Gebbie's house, I feel a little guilty about what I am planning to ask her. Melinda is an American artist and illustrator. She made her name in underground feminist comics in the 1970s before moving to Britain in the mid-1980s. After working on animated films such as *When the Wind Blows* (1986), she spent 16 years illustrating the highly sexual graphic novel *Lost Girls* (2006), written by her now-husband Alan Moore. Melinda and Alan describe *Lost Girls* as pornography, although censors and lawyers the world over disagree, and insist that it is art.

In 2015, Melinda returned to California for a few months and came back deeply disturbed by the impact of drought on her old home state. Having spent years away, the changes in the country-side were immediately apparent to her, yet the Californians she spoke to seemed oblivious. After her return she was so upset by the changes that she found the subject hard to talk about. This is why I feel bad asking her to describe it to me.

'You could see the difference straight away, like in the colour of the grass,' Melinda says. 'The California grass has always been a bit browny. It's always been a kind of army green, it's never been a fresh green like English green. I was told when I worked for a little history museum in the middle of the peninsula that the reason our grass was so poor was because it was Spanish grass, which the Spanish had planted for their cattle, and whatever local grass had been there was just covered up by this kind of army green, tufty stuff that wasn't much to look at. So the hills were always browny green, kind of olivey-looking in the best of times.'

This is how Melinda talks: in technicolour. She sees the world in terms of colours and emotions. Her world is more vivid and immediate than most, and if you spend time in her company you start to see the world that way too.

'When I went back in 2015, the grass was really pale yellow, like straw,' she continues. 'All the grass had turned yellowish-white. What really scared me was that, in Yosemite, the giant redwood trees were dying. And as far as I understand, the roots of a tree go as far down as the branches go up, so if it was that dry that these huge redwood trees were dying, that was really frightening. Yosemite is kind of a catchment area for Korean tourists – they like to see the redwood trees. But as they were walking round, they had no idea what a bad state it was in. They didn't realise that the waterfalls had dried up.

'People seemed absolutely oblivious. I never heard it discussed in any restaurants, on any trains, on any streets or ever on the news. The drought was just never mentioned. Except for the fact that they'd say that certain fires are happening in northern California, as if they'd always had fires, when they'd never really had so many fires before this drought.

'Nobody gave a damn. They were bathing whenever they felt like it, with big bathloads, or using the dishwasher for three people. They would see the animals suffering and they just didn't care. I remember seeing a buck – a massive deer with big horns, sitting and panting in the garden next door, desperate for water. All the animals were desperate for water. I left out bowls of water for the animals at night, and I would leave fruit, and I got bird seed to feed the blue jays. There were buzzards overhead, looking for the dead animals. The Californians that I saw didn't even seem to clock the fact that they were going into an incredibly serious drought. But it was already so severe! Their attitude really shocked and frightened me.'

After Melinda returned to the UK, California was hit by an unprecedented number of wildfires. The fires in the tinder-dry Napa Valley in 2017 were the most destructive in American history, killing 43 people and causing damage worth $18bn dollars. According to the science writer Andy Coghlan, the wildfires 'were so ferocious that they had the same effect on the planet as a volcanic eruption. The heat and smoke created a type of thunderstorm known as a pyrocumulonimbus, or pyroCb. The massive storm pumped the smoke from the fires so high in the atmosphere that it spread over the entire northern hemisphere.'

I recall seeing accounts of these Californian fires on the news, but in truth I hadn't really paid much attention to them. They seemed too distant to concern myself with. After hearing Melinda describe the drought, however, I downloaded a 360-degree VR video of the fires in order to gain a more visceral understanding of what they were like. With a head-mounted display strapped to my face, I immediately grasped the size of the fires and the scale of the landscape. I saw the way the ground smoked, how the sky glowed orange at night, and the exhaustion on the faces of the Californian firefighters. One 11-minute film allowed me to understand the reality of the wildfires in a way that countless news reports couldn't, such is the power of VR.

Melinda's account of her visit to California is sobering, because we've all noticed the increase in extreme weather events around the globe, from floods to droughts to storms of record-breaking intensity. You need to live in an extreme, counterfactual reality tunnel to be blind to the science of climate change. And yet there is still a belief that the problems are a long way off, and that we don't have to worry about them today. What Melinda's story reminds us is that people are prepared to ignore the reality of climate change, even as they are living through it.

2.

In 2009, the writers Paul Kingsnorth and Dougald Hine self-published a 20-page pamphlet with the title *Uncivilisation: The Dark Mountain Manifesto*, which immediately attracted considerable attention. Articles about it appeared in publications such as the *New York Times*, the *Guardian* and the *New Statesman*, which gave it a two-page lead review. The manifesto's argument was simple: it is too late to prevent the ongoing ecological collapse, civilisation as we know it is doomed, and we should stop fooling ourselves and come to terms with this.

According to the manifesto, a change in the circumambient mythos is both necessary and unavoidable. 'The myth of progress is to us what the myth of god-given warrior prowess was to the Romans, or the myth of eternal salvation was to the conquistadors: without it, our efforts cannot be sustained,' it tells us. 'Now a familiar human story is being played out. It is the story of an empire corroding from within. It is the story of a people who believed, for a long time, that their actions did not have consequences. It is the story of how that people will cope with the crumbling of their own myth. It is our story.'

The Dark Mountain Project, as the movement this manifesto

inspired has become known, has attracted a great number of dis-illusioned ecologists. These are people with a deep understanding of how interconnected our ecosystems are, who understand how the damage they are currently experiencing seems sufficient to bring down the whole system in its current form. To the supporters of the Dark Mountain Project, they are simply being realists. To their opponents, they are tired and bitter and are taking the easy option of giving up.

For environmentalists, there are plenty of reasons to find this convincing, not least of which is the horrifying collapse in bio-diversity and the increase in the rate of species extinction during the past 100 years or so. Biodiversity refers to the number of dif-ferent plant and animal species that share the same area, with a larger number of different species considered to be healthier for the ecosystem as a whole. Conservationists use the acronym HIPPO to refer to the causes of ecosystem degradation and species loss. It stands for Habitat destruction, Invasive species, Pollution, Population growth and Overhunting. The global efforts by con-servationists to save species from extinction have been heartening, but nowhere near enough to compensate for all the HIPPO factors.

If you can imagine that a wooden block is a species, then an ecosystem is a lot like the game Jenga. At the start of the game, the large number of blocks combine into a stable structure. All the weight of the individual blocks balances out, and this supports the larger whole. You then start to remove individual blocks, or species, from the larger structure. This causes the pressure from those blocks around and above the newly formed hole to shift, but the structure can usually absorb this change. Or, at least, it can in the beginning stages of the game. As more and more blocks go, the chances of total collapse increase. The impact of climate change on all this is like playing Jenga next to a busy motorway, with trucks rumbling past. It does not help, to put it mildly.

When collapse happens, it is usually caused by a complicated chain of linked events that ripple across the ecosystem. It could start with the loss of a keystone species such as a dominant predator. This would then cause an unsustainable increase in the populations of species previously kept in check by that predator. This in turn will exhaust their food supply, which would lead to a sudden die-off and the collapse of the previously balanced system.

The Dark Mountain movement's reaction to this situation is to focus on art, which is a means to rewrite the circumambient mythos and help the rest of us wake up to what is happening. As they describe themselves on their website, 'The Dark Mountain Project is a network of writers, artists and thinkers who have stopped believing the stories our civilisation tells itself. We see that the world is entering an age of ecological collapse, material contraction and social and political unravelling, and we want our cultural responses to reflect this reality rather than denying it. The Project grew out of a feeling that contemporary art and literature were failing to respond honestly or adequately to the scale of our entwined ecological, economic and social crises. We believe that writing and art have a crucial role to play in coming to terms with this reality, and in questioning its foundations.'

In the years since 2009, the Dark Mountain Project has produced a significant amount of work. They currently issue two beautifully produced hardback books a year, each full of poems, essays and artwork. I confess I have not been impressed by the content of these books. The original manifesto itself is focused, insightful and thought-provoking, but I've personally found the rest of their output to be a little humourless and self-righteous. As creative statements about the end of civilisation go, I feel that they fail to reach the artistic level achieved by *Game Over*, the 1986 debut album by the American thrash-metal band Nuclear Assault.

Statements about how we are all doomed get an easy ride in

our culture. We tend to treat negativity as serious and worthy of respect, and view positivity as suspicious and probably an attempt to sell something. Part of the reason for this, I suspect, is the funding models we have for news and advertising. News wants to grab our attention, and it does that by declaring terrible things are happening. Optimistic messages are more usually found in adverts, but these are known to be manipulative and economic with the truth. In these circumstances, standing up and declaring that something is terrible will gain you more credibility than declaring that something is wonderful.

There is also a strong, psychological desire to accept that we are all doomed, which seems to be especially prevalent in my generation. Once you adopt that position you no longer need to worry, or keep thinking about the problem, or look for solutions. Giving up is seductively easy, especially as you get older. The comedian Robin Ince has talked about how he used to constantly complain and moan about the world on social media, until he had a moment of clarity about what this was doing to him. If he continued on that path, he realised, he would ultimately turn into Morrissey.

I am unable to accept the Dark Mountain Project's certainty about the apocalypse. My opinion is biased here, because I have a great fondness for doubt. But making firm predictions about something as vast, vital and unpredictable as the world's political, economic and ecological networks seems to be a wild folly. While I am unable to agree that we are definitely doomed, I accept that if everybody thought as they did, and stopped trying to build a sustainable future, then we would indeed be stuffed. As a July 2018 *New Scientist* editorial put it, 'the biggest cause of biodiversity loss is inaction through despair'. We do need to actively find a way to halt the decline in biodiversity.

For all that I am tempted to dismiss the Dark Mountain Project as half a story, I am unable to dismiss the work of their co-founder

Paul Kingsnorth. Kingsnorth's novel *The Wake* is, for me, one of the finest novels of the twenty-first century. I do not see the world in the same way as he does, but I do consider him to be someone worth reading.

When I started reading Kingsnorth's wider writing, I was first struck by a level of pessimism that would make a teenage goth proud. 'Look: here's how it is, how it seems to me right now,' he wrote in *Dark Mountain*, issue 2. 'Life is a series of collapses, staggered and staggering. If there is a trick – and we seem to think there always ought to be – then maybe it is simply to remember that collapse is not always bad. Death is not always bad. Suicide: maybe even suicide is not always bad. Or if it is, if it is always irretrievably bad, at least maybe it is not always your fault.'

It is easy to dismiss this as self-pity or clinical depression, but if you continue to read Kingsnorth you realise that the reasons for his world view are much more interesting. Kingsnorth's sense of spirituality is indivisible from his love of wilderness. It is when humans are absent from the natural world that Kingsnorth feels the presence of the divine. As he wrote in 2014, 'One of the driving forces in my life is a deep love of nature. If you ask me to explain precisely what I mean by that, or why it has such a grip on me, I won't be able to. But I could tell you about profound experiences that I've had in forests and mountains, about the joy that rises in my heart when I see a hawk circle or hear the roar of an untamed river and the misery that sinks into it if I'm trapped in a city or on a motorway. I could tell you about the occasional brief glimpses I get into the reality that I am a passing moment in an ancient, beautiful, terrifying whorl of life on a vast unknowable planet; that I am not an observer of it, but a part of its wide flow, that there is no such thing as outside. This kind of thing is nearly impossible to put down on paper, as you can see. Once upon a time, many millennia ago, I suspect that it would have been the default world view, but today it is a hard one to live with.'

That such a world view is 'a hard one to live with' is a constant theme in Kingsnorth's writing. The increasing contamination of the planet's untouched wilderness, from this perspective, is blasphemous, but he is unable to prevent it. Not enough other people share his desire to preserve our remaining wildernesses. Most people view spoiling the natural world to be a fair price for our growth-based consumerist society. Kingsnorth's struggle, then, is how to make a life for himself in a society that actively progresses in a direction he sees as hell.

Within the environmental movement and conservative politics, there is often a desire to return to a lost 'golden age'. The date of this golden age varies quite widely. For many, it corresponds to their childhood years, before we had responsibility or an awareness of loss, when the world seemed full of wonder and, so far as we could tell, things generally made sense. Others would prefer to wind back the clock to before the Industrial Revolution, to live at a quieter place more in tune with the natural world. The author John Michell had a longing for a pre-Reformation medieval 'merry England'. His romantic view of this period seems entirely devoid of plagues, famines, medieval medicine and the worst excesses of the feudal system.

Others dream of going back further. They point to the moment when we started to practise agriculture, about 12,000 years ago, as the moment when humanity went wrong. There are good reasons to support this idea. Our hunter-gatherer ancestors were typically healthier and larger, and lived longer, than the first farmers. The amount of living matter on the planet has fallen by half since the beginning of human civilisation, mainly due to the clearing of forests for farmland. Settling down and planting crops meant that a given area of land produced considerably more food, which allowed population numbers to increase massively. This was a great boon for humanity as a whole, as it meant that we could live together in greater numbers and it allowed human culture to develop. But for individual farmers it also meant a life of disease, routine and back-breaking work. When Kingsnorth speaks of a time 'many millennia ago', when he believes his world view would have been humanity's default, it is tempting to locate this time back in our hunter-gatherer past, before we yoked animals and put them to work and before we began controlling the fauna and flora around us. When we dream back to such a time, it is tempting to imagine that humanity was then at one with nature, a small harmonious part of the ecosystem rather than something that stood apart from it.

This idea, unfortunately, doesn't hold water, because humanity was viciously out of sync with natural ecosystems long before we started farming. We have been ever since the cognitive revolution 70,000 years ago, when we developed the ability to co-operate using shared myths and fictions. When we headed out of Africa, we disrupted every ecosystem we came across.

We arrived in Australia about 45,000 years ago, which at that time was home to extraordinary megafauna. There were wombat-like animals called Diprotodons that were about the size of hippo-potamuses. There were flightless birds who were three metres tall, and giant carnivorous lizards that were up to seven metres in length. But shortly after humans arrived, the great majority of these Australian megafauna became extinct. We arrived in the Americas perhaps 16,000 years ago, at which point the American megafauna also went extinct. When *Homo sapiens* encountered other human species such as Neanderthals or *Homo floresiensis*, they too went extinct. It's hard to deny that we seem to be the common link in all these extinctions.

If you want to find a time when our ancestors were not

destroying ecosystems, you need to go all the way back to the time before the cognitive revolution. This was when we changed from being just like every other animal species to having modern-style minds which created their own reality tunnels and began conceptualising our world. You would not be able to appreciate nature in the way Kingsnorth does before this, because you would have no concept of it, in the same way that a fish has no concept of water. Before the cognitive revolution, we were about as spiritually aware as the average horse. Some might say that this would be no bad thing, of course, but the desire to become like animals again is a very niche position. Personally, I would be horrified by the thought of losing our immaterial selves, with all the art, philosophy, stories and humour that they contain.

I suspect this is why Kingsnorth calls his world view 'a hard one to live with'. He seems to know that there is no way he can satisfy his yearning desire for wilderness. The awareness of what makes nature sacred is what separates us from it. There is no way to square this circle. It is like the grief felt at the loss of a loved one, in that there is no way to make it right. It can only be lived with, and this will be hard. The Dark Mountain Project, if seen in these terms, is hard to criticise. Giving up eventually becomes inevitable when what you seek can never be found.

3.

Back in medieval times, the circumambient mythos centred on a hierarchy called 'the great chain of being'. At the top of this, naturally enough, sat God. Below God came the archangels, angels and other spiritual beings. Underneath these came mankind, with kings and princes at the top, and commoners and slaves at the bottom. Below mankind came wild beasts, followed by domestic animals, trees, the rest of the plant kingdom, gemstones, precious metals and, at the bottom, the rest of the minerals and rocks of the planet. This great chain of being was a guide to authority and importance. Things higher up had the right to do what they wanted with the things beneath them.

When the medieval world gave way to the Enlightenment, philosophers like Immanuel Kant showed that it was possible to talk of morality without involving God. Kant constructed a theory of morality that rested on human reason and which did not need God at all. It is thanks to people like Kant that we now have the concept of human rights, and that the American Declaration of Independence states that 'We hold these truths to be self-evident, that all men are created equal.'

It is sometimes said that the Enlightenment broke the great chain of being, but it is probably more accurate to say that it just knocked the top part of it off. After the Enlightenment we no longer had to invoke the authority of God in politics or bureaucracy, but mankind's place above the animal, plant and mineral kingdoms remained unchanged. We still had the right to do whatever we wanted with those. Under the new mythos of growth and progress, exploiting the natural world for humanity's sake was believed admirable. To be fair to Kant and his followers, this didn't seem problematic back when he was writing, towards the end of the eighteenth century. The population of the world then was about 800 million. Nowadays, the world's population is over 7 billion. That does change things.

As members of the Dark Mountain Project would be the first to tell you, exploiting the modern world at the same rate that we are currently doing is in no way sustainable. To give some unhappy examples, the Intergovernmental Panel on Biodiversity and Eco-system Services tells us that 37 per cent of Europe's freshwater fish face extinction by 2100, along with half of all Africa's birds and mammals, and that exploitable fish stocks off Asia–Pacific coastlines will collapse by 2048. A study into the decline of animal populations, published in the peer-reviewed journal *Proceedings of the National Academy of Sciences*, claimed that the loss of populations of mammals, birds, reptiles and amphibians all over the planet had to be counted in billions. A mass extinction event is ongoing, the study argued, and blamed 'human overpopulation and continued population growth, and overconsumption, especially by the rich'. As the naturalist Michael McCarthy has pointed out, 'Most Britons remain blithely unaware that since the Beatles broke up, we have wiped out half our wildlife.' Perhaps the most shocking statistic comes from a 2015 UN report. This claimed that, if soil depletion continued at its current rate, then the world had only 60 harvests left.

Population is expected to keep growing, with the UN predicting that it will reach 11.2 billion by the end of the twenty-first century and continue to increase further in the twenty-second century. As the mass extinction study's authors stress, however, the numbers involved in population are not the full story. More problematic is the percentage of resources that are consumed by the richest. If there is to be any hope of avoiding a systematic collapse, then the level of consumption aspired to by us in the West must decline. The problem is that Western culture is engineered to makes us want stuff. It is designed to keep stimulating the D1 receptors in our brain. As the truism goes, most people find it easier to imagine the end of the world than to imagine the end of capitalism.

To avoid the worst of the ecological problems that are being stored up, our culture and economy are going to have to behave very differently to what is now the norm. This looks horribly unlikely if you assume that people will always think in the same way that people from the twentieth century do. Fortunately, this is not the case. The post-Millennial generation, who think in terms of networks rather than isolated individualism and who find social features such as Snapchat streaks more addictive than sharing fake news on Facebook, are already very different to Baby Boomers, Gen Xers and Millennials. This is a generation who appear to respond more to having their social D2 receptors stimulated than their needy D1s. Or, to put it another way, these young people are exactly what the world needs right now.

Some caution is needed here. When big-picture generalisations about generations are compared, the focus naturally falls on the differences rather than the similarities. We view punks as radically different to the hippies that preceded them, for example, but the two cultures have far more in common than they would like to admit. Focusing on differences can lead us to think that each new generation is a radically different species to the one that came before when, ultimately, they are always more alike than different. They all want friends, and security, and excitement. Not every member of a generation matches the generational trends that emerge from the data. Not every post-Millennial is deeply empathetic or has anxiety issues. At best, the differences that we focus on point to a change in direction, rather than a mass migration. Children being born now are certainly going to experience a lot of activity in their needy D1 receptors, just by growing up in the culture created by everyone from the Boomers to the Millennials.

The generational change is welcome, but it does not seem to be enough to bring about a sustainable future by itself. The problem is that our circumambient mythos still has the great chain of being buried at its heart. For as long as we believe that we have the right to exploit the natural world as much as we want, attempts to rein in our use of the natural world to sustainable levels will be interpreted as some form of painful restraint. For all that our intellect argues in favour of the necessity of limits, our emotions will always fight against going without. What is needed is an idea that replaces the great chain of being at the heart of our ideologies. If such an idea is to have a hope of being accepted, it would have to be simple, fair and self-evident.

As it happens, the American biologist E. O. Wilson has had just that sort of idea.

4.

Wilson's proposal is called Half-Earth. The idea is that half the world is used and exploited by humans, while the other half is given back to nature. 'Only by committing half of the planet's surface to nature can we hope to save the immensity of life-forms that compose it,' Wilson argues. 'A biographical scan of Earth's principal habitats shows that a full representation of its ecosystems and the vast majority of its species can be saved within half the planet's surface. At one-half and above, life on Earth enters the safe zone. Within half, existing calculations from existing ecosystems indicate that more than 80 percent of the species would be stabilized.'

This, clearly, is quite an ambition. Wilson's figure of '50 per cent' may at first seem suspiciously neat, but other environmental scientists agree that it is roughly correct. 'Targets like 50 per cent are in the right ball park when it comes to the minimal amount of area needed to conserve biodiversity,' says James Watson at the University of Queensland. Jonathan Baillie of the National Geographic Society believes the issue is time-critical, and that we need to hit the 50 per cent mark by 2050.

The great majority of the world's nations now have national parks or government-protected wildlife reserves. This practice started with the American creation of Yellowstone National Park in 1872 and has continued to this day. Since the South Downs National Park was established in England in 2011, I've been lucky enough to live just two and a half miles from a national park. The problem is that the amount of the Earth's surface which is protected in this way is about 15 per cent, along with 7 per cent of our oceans. Targets set by the Convention on Biological Diversity require countries to turn 17 per cent of their land into protected areas by 2020. We are moving in the right direction, but there is a long way to go to get up to Wilson's 50 per cent. It does not look an easy task at a time of rapid population growth and increasing demand for resources. Nevertheless, we may be hasty in dismissing the idea out of hand.

No one is arguing that forcibly moving people off their land is justifiable. History has enough examples, such as the forced land clearance by Scottish landowners in the late eighteenth century, to teach us how cruel and destructive such actions can be. But movement from rural areas to cities is occurring naturally. As the author Nassim Nicholas Taleb advises, and as the young seem to intuitively realise, the best way to make extraordinary things happen in your life is to live in cities and go to parties. It is increasingly common for young people in isolated villages to move away in search of work, leaving behind dwindling, ageing and increasingly unviable communities. Increased competition brought about by globalisation and a collapse of farm commodity prices are blamed for a widespread withdrawal from marginal farming land in Europe. A 2014 report by Rewilding Europe claims that, by 2030, more than 116,000 square miles of farmland will be abandoned, which is 5 million hectares larger than the size of Britain, and that by 2020 four out of five Europeans will be living in urban areas. If this trend towards city life is happening regardless, it could be further encouraged by policies such as a land value tax, or by building regulations that favour greener city apartments over sprawling suburbs. As the South Downs National Park shows, it is possible to protect natural environments in collaboration with the people who already live and work in those environments, rather than requiring the complete abandonment of those communities. The idea of gradually expanding existing protected areas is generally not controversial, and usually is welcomed.

As I was writing this chapter, Colombia's Serranía de Chiribiquete tropical rainforest national park was declared a UNESCO World Heritage site and increased to 4.3 million hectares, making it the world's largest protected rainforest. The Seychelles government also announced it was creating two new marine protected areas, which will cover more than 81,000 square miles, an area about the size of Britain. These stories did not gain much attention in the press, as there is not much of a market for positive environmental news, but they are examples of how many countries are putting space aside for nature. Trees covered 7 per cent of England in 1980, for example, but that has since risen to 8.4 per cent and is increasing rapidly. According to researchers at the University of Helsinki, between 1990 and 2015 forests grew by 1.3 per cent in high-income countries and by 0.5 per cent in higher-middle-income countries. Unfortunately, they also shrank by 0.3 per cent in lower-middle-income countries and by 0.7 per cent in poor ones, a reminder of the important role that wealth plays in conservation.

The Half-Earth idea is appealing because it rewrites the narrative of conservation, turning it from a tragedy to a quest. One problem the conservation movement has is that it is understood in military terms. It is a struggle and a fight, where the occasional victory is drowned out by the overwhelming tide of defeats. It is a story of heart-breaking loss, where even if you may win the battle you are going to lose the war. Half-Earth turns this on its head. It takes something which is currently a reactive process and turns it into proactive achievements. The gradual increase of land given over to nature is a positive story of incremental progress towards a clear, easily understood goal. Every extra 1 per cent is reason to celebrate. Giving 50 per cent of the world over to nature will not happen in the immediate future, but as a long-term goal reached incrementally by countless tiny steps in the right direction, it is a goal we are already travelling towards.

The real value of the Half-Earth idea is that it does not tweak the surface details of the circumambient mythos, but instead it corrects the flaw at its heart. It declares that nature and humanity are equally important. In doing so, Half-Earth removes the remains of the great chain of being, which Kant left in place. The importance of this cannot be overemphasised, because the great chain of being is so central to our understanding of the world. While few people may have heard of the great chain, our ideologies and business practices show that pretty much everybody has internalised the idea. As Wilson describes mankind, we are currently 'Obsequious to imagined higher beings, contemptuous towards lower forms of life.' Half-Earth changes this, defining us as nature's equal. Digitally enhanced mankind and nature are the dual inhabitants of Earth. They are not slave and master, but instead must try to live together in a sustainable partnership.

To achieve Half-Earth would be a huge undertaking. It would take planning by both national and international bodies. Different targets for each country would need to be defined and areas for potential new nature reserves identified. These protected nature reserves would then be managed with the goal of increasing biodiversity, for example by reintroducing lost keystone species or removing invasive species in a process called 'rewilding'. Rewilding projects in Europe have already led to the successful reintroduction of wolves in Germany, the Iberian lynx in Spain, elk in Denmark and bison in the Netherlands. Wherever possible, natural 'corridors' would be created to link together protected reserves, because connected ecosystems have greater biodiversity than small islands of nature. If these reserves could grow to the required size, and if – it's a big if – the planet continues to de-carbonise its power systems as mandated by the Paris Agreement, then there would no longer be a desperate need to save the planet. The planet would quietly heal itself.

5.

Just north of the South Downs National Park is a 3,500-acre estate called Knepp. This was marginal farming land, with heavy clay soils, and attempts to farm it commercially pushed its owners, the Burrell family, close to bankruptcy. In the year 2000 they decided to give up intensive farming and turn the land over to the natural processes of a Sussex ecosystem. Red, roe and fallow deer, Old English longhorn cattle, Tamworth pigs and Exmoor ponies were introduced to the land and left to their own devices. The spraying of fungicides, pesticides and artificial fertilisers stopped. The Burrells then sat back and interfered as little as the law allowed. Neighbouring farmers were horrified. To abandon intensive farming like this was seen as a dereliction of care by landowners.

The explosion of biodiversity that has erupted at Knepp since then has been nothing short of astonishing. The air is thick with birdsong and insects, and colour and variety assault the eyes. Most attention has been focused on the arrival of rare and threatened species, such as nightingales, purple emperor butterflies and turtle doves. These were once synonymous with the British countryside but are now close to extinction. Yet it is the richness of all eco-systems across all scales, not the arrival of individual species, that is important. The way that dung beetles, earthworms, fungi and orchids are thriving is every bit as important as the fact that all five species of British owl, or 13 of the 17 UK bat species, can now be found there. Knepp is a textbook example of how interconnected the natural world is. If you wanted to save or protect the night-ingale or the purple emperor butterfly by providing all that they might want in a protected area, you would almost certainly fail. The Burrells did not set out to attract or protect these species. They allowed the natural ecosystem to live, and nature did the rest.

Knepp has shown us that a lot of what we think we know about British wildlife is wrong. Numerous species have been found thriving in Knepp's scrubland that were previously believed to be woodland dwellers. It turns out that the reason we kept finding these species in woodland was because their preferred habitats had been destroyed. The key to creating a vibrant, natural ecosystem, it seems, are the actions of grazing ruminants such as cattle or dear, as well as snuffling and digging creatures like pigs, who prevent the landscape from becoming close-canopy forest. The presence of these larger animals creates the conditions that keep the smaller ecosystems of small mammals, insects, amphibians and even soil bacteria healthy. We are only now starting to understand the extent to which healthy soils remove carbon from the atmosphere, and how much soil can help in the fight against climate change.

Knepp is not a complete natural ecosystem. It lacks a natural predator to keep the deer and cattle numbers down. The ideal animal, in this landscape, would be a lynx, although this would not go down well with Sussex dog-walkers. Instead, animals are culled when their numbers demand it, and their meat is sold. It is also not legally possible to allow dead cattle or horses to decom-pose naturally when they die, even though the processes involved would play an important role in the ecosystem. But despite these constraints, Knepp is wild enough to reveal to us the true state of the British countryside. As Knepp's co-owner, the writer Isabella Tree, has said, 'Rewilding Knepp has changed the way we look at the world and much of it is depressing. When we go for a walk with friends elsewhere in the countryside – the same walks we used to enjoy without thinking in the past – chances are what we notice most are the silence and the stillness. As the landscape flashes by on a train or motorway, we now know what isn't there. Compared with Knepp, most of Britain seems like a desert.' But thanks to places like Knepp, we are now aware of the problem, and aware of what a natural ecosystem should look like. We also know how quickly ecosystems can be reinvigorated.

Rewilding is not easy. Ecosystems are such complex and inter-related networks that great care has to be taken in deciding which animals to add. It can be controversial, as it was when a Dutch scheme to rewild marshland east of Amsterdam resulted in mass deaths of deer, horses and cattle following the harsh winter of 2017–18. It's worth noting, though, that rewilding ecosystems is exactly the sort of complex problem which AI is ideally suited to tackling. Scientists at Imperial College London have begun using solar-powered listening devices to monitor huge tracts of Borneo rainforest, for example. These are built from cheap Raspberry Pi computers which upload the sounds they hear to a remote server. AI then listens to these recordings and identifies which animals are present by their calls. In this way, the populations of huge numbers of species can be monitored over time without human intervention, and the factors that are causing biodiversity to increase or decrease can be identified.

After rewilding, it is necessary to ensure controlled access to these areas. This is vital because, as Wilson remarks, 'It has been my impression that those most uncaring and prone to be dismissive of the wildlands and the magnificent biodiversity these lands still shelter are quite often the same people who have had the least personal experience of either.' Nature needs to be undisturbed, but that doesn't mean we should not spend time with it. Digital urban living is only part of the story of who we are. To quote the pioneering Prussian naturalist Alexander von Humboldt, 'The most dangerous world view is the world view of those who have not viewed the world.'

This is the reason I visit Knepp at dawn on a cold October morning. A thick mist covers the ground, veiling much of the fauna and flora. But very quickly I notice differences between Knepp and other local landscapes. Fallen branches lie untouched beneath the trees that shed them. I see a large longhorn cow lying with a calf in a small area of woodland, rather than standing in an open field with the rest of the herd as I would have expected. When I see a swan ungainly waddling across open landscape some distance from water, I am reminded of how much our expectations about where animals should be is based on where we allow them to be.

I have come here to witness the annual deer rut. This is how male deer prove their fitness and standing in their social hierarchy, in order to woo the females, and it takes place for a few weeks every autumn. The thick, sickly smell of pheromones hangs in the early-morning mist, pushing deer-kind out of their normal behaviour patterns and towards violence and sex. The deer rut is like a cross between *Bambi* and *The Purge*. The ungodly bellowing of the bucks carries through the cold air and sounds like pigs attempting a Maori haka.

Knepp runs safaris at dawn and dusk so that visitors can witness the deer rut. Safaris like this are part of how the estate works financially, and the deer-rut expeditions all sell out long in advance. Along with 12 other visitors, I am driven around the estate in an open-sided Pinzgauer military vehicle, peering through the mist for the bellowing deer.

We hear the clack of antlers before we see the fighting animals. Two fallow bucks are engaged in ritual combat not far from where we park, on the other side of a small pond. I had not been prepared for how ritualistic a fight between two beasts could be. They are attempting to kill each other by gouging their antlers into their opponent's flank, but these violent clashes are interspersed with graceful, controlled behaviour called 'parallel walking'. This is, as the name suggests, when two deer walk alongside each other with their heads forward, in a calm, seemingly carefree manner. In truth, this is a test of nerves and bravery, as each deer is waiting for the other to turn its head and attempt to gouge it open with its antlers. The moment one deer turns, the other matches it with lightning reflexes and blocks the attack with its own antlers. After clashing heads like this, the calm, stately parallel walking continues again, as if nothing untoward or violent had happened. This ritual continues until one deer accepts it is weaker and runs away in defeat, which is the most common result, or until one animal kills the other. I have the strange feeling that I am witnessing the origins of jousting. For all that the code of chivalry justified medieval jousts, it really wasn't that different to animal behaviour.

My visit followed the endless heatwave of summer 2018. I had wondered what impact that would have on this unmanaged ecosystem, but Knepp, it turns out, was relatively untroubled by the extreme weather. Now that previous attempts to drain the land for agriculture have ended, and drainage ditches have silted up, the land has regained its natural ability to retain water. Even during the extreme weather that scorched the land elsewhere, the grass here remained green.

Before visiting Knepp, I had unconsciously come to view nature as a weak, sickly thing that would perish if we did not nurse it back to health. Now, I realise I had that entirely wrong. Nature is wild and powerful, and it will roar back to life like a bellowing buck the moment we step away and allow it to do so. There are 26,000 species that are threatened with extinction, according to the International Union for Conservation of Nature red list. Every time a species becomes extinct it is a tragedy, for we will never see their like again. But those 26,000 are 0.3 per cent of the 8.7

million species on Earth. Even if the rest of these species are under pressure and their numbers are dwindling, they will return with force the moment they get the opportunity.

What's important about the rewilded Half-Earth goal is not just that it's the only viable relationship between mankind and nature on the table. It's the fact that it is on the table. It is an option we could take. The common assumption that we are all doomed and that there is nothing we can do starts to look a little daft when there is an idea out there which we could work towards.

No one is saying that it will be easy. No one is saying that we are certain to succeed. But it sure beats giving up.

6.

In a fit of optimism, I attempted to clear out the bedroom cup-boards. Among the accumulated clutter I found a box of abandoned gadgets and pieces of technology, which were about 10–15 years old. The box contained a CD Walkman, early digital cameras, non-digital cameras, a Dictaphone, an early 'personal digital assistant' called the Handspring Visor, which came with a stylus and handwriting recognition, a couple of modems, an old mobile phone and an early MP3 player, together with an assortment of product-specific cables and chargers. None of this was cheap to buy at the time, so it wasn't thrown away when it became redundant. Instead, it was carefully stored away for years, until the day finally came when it was rediscovered in the back of the wardrobe. Then it was thrown away. Or rather, it was taken to the correct part of the local recycling centre, as is now the done thing.

What struck me about all that kit was how unnecessary it now was. I don't need any of these devices any more because my phone does everything all that tech did, and it does it considerably better.

It's not even a new phone. I've had it for over two years, the age by which an early smartphone would have been ready to be replaced. But I have no need to upgrade, for the phone works well and newer models don't offer anything else that I have any need for. I find myself in a situation that is novel in my Generation X life, which is that for the first time since my early teenage years I am not hankering for a new piece of technology. So long as I have my phone, laptop and a decent speaker or two, then my needs are met.

I am not alone in no longer needing more stuff. The average Briton used 15.1 metric tonnes of material per person in 2001, but by 2013 that had reduced by almost a third to 10.3 metric tonnes, and the trend looks set to continue. An increasing share of the economy is being driven by businesses that deal in code and virtual goods, from bitcoin miners to videogames and streaming services such as Spotify or Netflix. Uncles and aunties across the land are getting used to the idea that what their nieces and nephews really want as Christmas or birthday presents are V-Bucks, Steam vouchers or some other form of digital currency exchange. Generation Z also have less interest in learning to drive than older generations, as the expense of owning a car is less appealing in the age of Uber and ride-sharing apps. The cultural shift from valuing possessions to valuing skills, connections and experiences has been much discussed in the lifestyle pages of broadsheet newspapers, and it fits nicely with the more empathetic, emotional values of the young. An industry has grown up around the practice of decluttering. Having stuff is no longer the status symbol that it once was.

That doesn't mean that our environmental problems are over, unfortunately. Things may have turned a corner and are moving in the right direction, but we're still using too much stuff, and our economic and political systems continue to encourage us to consume more. Even environmentalists who are more positive than the Dark Mountain Project fear it will take a radical change to our economic system before we can live within the carrying capacity of our planet's ecosystems.

If you're searching for a radical alternative political and economic system, there are many on offer. There are countless ideas available for different ways we could arrange our affairs. Some say a maximum voting age is all that is needed, because removing the vote from people when they collect their pensions would remove a demographic inclination to short-term planning. Others think that negative interest rates are what we need, as they would make the value of established wealth dwindle and encourage investment and infrastructure maintenance. A carbon tax on meat, to encourage vegetarian or vegan diets, could potentially have a significant impact on the climate. A digital version of Athenian democracy, under which political parties would be replaced by online voting by all citizens, has supporters among those concerned about political lobbying by corporations. A radical overhaul of the taxation system, taxing carbon use instead of income, might fix a few problems. A maximum-wealth law, under which people who become too rich are jailed, would be extremely controversial but has a certain cheeky charm. None of these ideas feel in any way likely, however. These are very niche positions which do not have widespread political support. It is hard to imagine the chain of events that would lead to systems like those being put into place.

An exception to this is an idea called Universal Basic Income. Like those other suggestions, it is undoubtedly a radical idea. With Basic Income, every single person is given a guaranteed income, sufficient to feed and house them, and they are then free to earn what they can on top of that. This replaces most existing welfare payments and is paid for by general taxation. Those who wanted a better standard of life could work as much as before, but falling out of the workforce at times would not have the devastating impact it now does.

8. FIXING THINGS

Unlike other radical ideas, there is a strong swell of support for Basic Income from around the world and from across the political spectrum. Many thinktanks and pressure groups are actively researching and refining how it might be implemented. Pilot schemes have been run to assess how different aspects of the idea play out in real-world situations, in countries as diverse as Finland, Kenya and Canada. It is championed on the political left as a means to reduce poverty and increase social mobility, and it is championed on the right as a way to increase entrepreneurship and self-reliance. It is not difficult to find leaders in places as different as Silicon Valley and the halls of Westminster who consider the idea not just plausible, but inevitable.

Much of the modern support for Basic Income is a reaction to predictions about the disappearance of good jobs in the coming technological future, and because the great majority of jobs that are being created now are low paid and insecure. Technology has always replaced jobs, of course, but it usually created better jobs in return. The invention of the motor car was not good news for stable boys and blacksmiths, but if they were willing to retrain as mechanics, engineers or drivers they had a more prosperous future ahead of them than in the age of the horse. The fear is that while digital technology is destroying good jobs, like other new technologies before it, it is not creating better jobs in sufficient numbers. AI also makes the problem worse. New technology not only has to create more jobs, but it has to create jobs that couldn't also be done cheaper and quicker by AI.

Once a sizeable percentage of the population is no longer part of the workplace, the problem is not just how those people will survive. The problem is also, who will buy the consumer goods and services that the rest of the economy is pumping out? This is what makes Basic Income attractive to economists. Paying every citizen a guaranteed amount, in this scenario, is a way to keep the whole show on the road.

There are a number of problems that hold back the adoption of Basic Income. A key issue is what I call the 'splutter test'. This test assesses whether or not you can describe the idea to readers of mass-market newspapers without causing them to accidentally spit tea all over you.

At the moment, the idea of free money for everyone tends to fail the splutter test. In the austerity years following the financial crisis of 2008, a narrative of 'strivers and skivers' has been employed by politicians and journalists to justify cuts. The notion of money going to someone who hasn't worked for it has been portrayed as a great wrong, and it will not be easy to reframe it as a great right.

A second problem with Basic Income is the name. It's an incredibly boring-sounding name. This makes it almost impossible to talk about Basic Income without sounding deathly dull. 'Basic' is not an inspiring word. There have been other names suggested over the years, such as 'citizen's income', 'negative income tax', 'life pension' or 'universal demogrant', but these have been equally uninspiring. My favourite name is the Stipe, as suggested by the writer Alistair Fruish. This is both a reference to a stipend and a nod to Michael Stipe from the band R.E.M., because the system would be *Automatic for the People*. Realistically, though, this system is now known as Basic Income and we will have to get used to that. If it helps, the Beatles, the euro and Doctor Who are also terrible names, but as soon as they become sufficiently familiar you no longer notice.

The biggest problem, at least when people first hear about the idea, is the expense. It will not be cheap to give everyone enough money to live on. When you first encounter the idea the natural inclination is to recoil and assume it would be so unaffordable that the whole concept needs to be dismissed as absurd. Regardless of how you designed it, a Basic Income for a moderately populated country would cost its government many billions of dollars each year.

It is not the case, however, that supporters of the idea are naive and wild-eyed and have not noticed this cost. Almost everyone who supports the idea initially assumed that it would be implausibly expensive but came to view it as financially possible when they looked into it further. A list of people who have supported the idea of a Basic Income includes Mark Zuckerberg, Archbishop Desmond Tutu, Tim Berners-Lee, Elon Musk, Caroline Lucas, Thomas Paine, Bertrand Russell, Richard Nixon, Virginia Woolf, William Morris, Nicola Sturgeon, Ray Kurzweil, Milton Friedman, Richard Branson, Robert Anton Wilson and Martin Luther King, Jr. This is clearly not a list of people who were too dumb to notice how expensive it would be. It is a list of people who have looked into the idea and who have come to the conclusion, despite their very different politics, priorities and world views, that it is economically plausible.

There are many factors that take the sting out of the cost. Basic Income replaces a large part of the existing welfare system, and it makes the bureaucracies employed to decide who is eligible for benefits unnecessary. There is the financial boost that comes from keeping everybody active in the economy, instead of losing a growing percentage to the 'economically inactive' category. Another economic benefit is an increase in entrepreneurship, and there is growing evidence for a range of secondary financial benefits, such as a reduction in mental- and physical-health-care costs, especially connected to conditions exacerbated by stress. Families would be in a better position to care for ill or elderly relatives, reducing the need for the state to provide social care.

Some claim that all of this would match the cost of the system, making Basic Income 'revenue neutral'. This seems optimistic, but it does suggest that the expense is not as absurdly implausible as first assumed. In most countries, there would need to be a further form of progressive taxation, perhaps as a form of VAT, land value tax or higher bands of income tax. As a result, implementing a Basic Income would be a major political change, in much the same way that implementing the National Health Service was. It is the sort of thing that is easy to dismiss as impossible and ruinously expensive, until somebody goes ahead and does it.

Ultimately, the biggest question surrounding Basic Income is not the expense, its dull name or the splutter test, it is something more fundamental. It is the question of what constitutes a good life.

7.

The Conservative MP for Grantham, Nick Boles, has argued that 'the main objection to the idea of a universal basic income is not practical but moral. Its enthusiasts suggest that when intelligent machines make most of us redundant, we will all dispense with the idea of earning a living and find true fulfilment in writing poetry, playing music and nurturing plants. That is dangerous nonsense. Mankind is hard-wired to work. We gain satisfaction from it. It gives us a sense of identity, purpose and belonging [. . .] we should not be trying to create a world in which most people do not feel the need to work.'

This is the opposite position to the more utopian perspective favoured by most Basic Income supporters. Their attitude is nicely summed up by the American inventor and systems theorist Buckminster Fuller, who in 1970 argued that 'We must do away with the absolutely specious notion that everybody has to earn a living [. . .] We keep inventing jobs because of this false idea that everybody has to be employed at some kind of drudgery because, according to Malthusian–Darwinian theory, he must justify his right to exist. So we have inspectors of inspectors and people making instruments for inspectors to inspect inspectors. The true business of people should be to go back to school and think about whatever it was they were thinking about before somebody came along and told them they had to earn a living.'

Many Basic Income supporters dismissed Boles's comments because of his background. Boles is from a very privileged section of society not known for its insight into the daily lives of the average working man or woman. But it is unfair to use Boles's background as a reason to dismiss his comments, because the issue he raises is worth looking at in more detail. Is mankind 'hard-wired to work'? Is that what gives us our 'sense of identity, purpose and belonging'? This is indeed a moral issue, as Boles points out. It is concerned with the question of what it is to live a good life. It raises issues concerning meaning, purpose and the nature of existence. These are subjects that economists generally keep their distance from.

A useful starting place for understanding such things is the work of the Austrian psychiatrist Victor E. Frankl. Frankl was a survivor of Auschwitz and other Nazi concentration camps, and later wrote the Holocaust memoir *Man's Search for Meaning*. According to Frankl, whether or not a concentration camp inmate lived or died depended on their attitude to the future. Prisoners who shared what crumbs of food they had, even though they were starving themselves, were more likely to make it through to the end of the war. As he wrote, 'The prisoner who had lost faith in the future – his future – was doomed. With his loss of belief in the future, he also lost his spiritual hold; he let himself decline and became subject to mental and physical decay. Usually this happened quite suddenly, in the form of a crisis, the symptoms of which were familiar to the experienced camp inmate.'

This collapse would usually be preceded by a preoccupation with the past, which served to make the present seem less real or important. When the loss of hope occurred, it was not unusual for an inmate to lie down and refuse to move. 'No entreaties, no blows, no threats had any effect,' Frankl recalled. 'He simply gave up. There he remained, lying in his own excreta, and nothing bothered him any more.'

In contrast, prisoners who could maintain a sense of purpose were more likely to physically survive the ordeal because, as Nietzsche once wrote, 'he who has a *why* to live for can bear almost any *how*.' For Frankl, it was the thought of seeing his wife again that kept him going. He did not know that his wife had already died of typhus in the Bergen–Belsen concentration camp. Even though the future he was living for could never happen, his love for his wife saved his life regardless.

Love is one reason for living but, as Frankl argues, there are others. A spiritual or religious belief is a common motivator, as is a career or vocation, freely chosen, that is believed to be worthwhile. After his experience in Nazi concentration camps, Frankl added the endurance of intense suffering to the list, although hopefully this is not one that many of us will need to cling to. It did not matter, Frankl learnt, what gave your life purpose, or what your route to meaning actually was, just so long as you had *something*. He compared asking what the meaning of life was to asking a chess grandmaster what the best move was. There was no one correct answer. Any answer given would depend upon individual circumstances. Frankl thought that the tasks that life presents us with, 'and therefore the meaning of life, differ from man to man, and from moment to moment. Thus it is impossible to define the meaning of life in a general way. Questions about the meaning of life can never be answered by sweeping statements.'

To Generation X, in particular, knowing that there was no

definitive, unarguable single answer to the meaning of life fed the belief that life was meaningless. In the 1980s, there was a considerable body of opinion that thought accumulating money was the closest thing we could come up with to a meaning of life. It was not an ideal solution, clearly, but it was the best we could do. The idea meshed with the Protestant work ethic that viewed hard work as a moral good in itself, as it was well known that the devil found work for idle hands. Nick Boles MP is the correct age and demographic, therefore, to find the idea that paid employment 'gives us a sense of identity, purpose and belonging' to be a convincing one. But how well does this idea stand up for more recent generations?

The Millennial generation that followed had a far healthier attitude to the question of meaning than Generation X. It did not trouble them that there was no single, universally agreed, meaning of life. Like Frankl, they were comfortable with the idea that purpose was something that we all had to find for ourselves. If something was experienced as personally meaningful, then by definition there was meaning to be found. To the Millennial generation, Frankl's belief that purpose could be found in a variety of different things for different people, such as love, spirit, vocation, suffering or a personal cocktail of those ingredients, seemed self-evident and valid in a way that it didn't to nihilistic Generation X or absolutist Baby Boomers.

The percentage of the young who found their sense of purpose solely in employment, however, has been declining as the quality of jobs available to the young falls. Millennials and Generation Z are increasingly likely to find themselves in the insecure gig economy rather than the steady, stable careers with good pensions that Baby Boomers enjoyed. A 2015 YouGov poll found that 37 per cent of British workers think that their job does not make a meaningful contribution to the world, and a 2013 Harvard Business

Review survey of 12,000 professionals found that half thought their job had no 'meaning and significance'.

The problem is what the anthropologist David Graeber calls 'bullshit jobs'. There are still jobs in fields such as nursing, teaching, plumbing or policework that are self-evidently worthwhile, but they are increasingly a minority. Three quarters of the jobs in the developed world are in admin or the service industries, and these are considerably less likely to provide you with a reason to get out of bed each morning. According to a University of Manchester study, while having a good job is better for your health than being out of work, unemployment is actually better for you than a stressful, badly paid, low-quality job. Boles is right when he argues that work can provide 'a sense of identity, purpose and belonging', but in the current job market it can't do that for everybody. If he is truly concerned about people finding a sense of meaning that can make their lives worthwhile, he should reconsider whether everybody is going to find that in the modern world of work.

A lot of this hinges on the definition of the word 'work'. In the twentieth century, a time of labour movements and unionisation, work was understood as paid employment. Operating a production line in a car factory or serving drinks behind a bar were work. Cooking and cleaning in a household, caring for the infirm or volunteering for local good causes were not. The actual effort needed to do these things may have been the same, but if there was no pay cheque at the end of the week then it didn't count as the type of work that gave meaning to life. Part of the appeal of Basic Income is that this definition is no longer relevant. Non-paying work can be recognised as a meaningful part of people's lives.

As well as work – however that word is defined – there is the question of the things we are drawn to do, and how much further we can explore them. These can be similar to the 'writing poetry, playing music and nurturing plants' that Boles was dismissive of.

I myself could spend a life writing, listening to music and nurturing plants and consider that to be a life well lived, but I know that this would not appeal to everyone. Others might be drawn to mastering a sport, founding a start-up, adopting stray animals, researching local history, climbing mountains, fostering, designing clothes, acting, hang gliding, rewilding, political campaigning, running marathons, learning languages, mentoring, sound engineering or building VR worlds in a way that does not interest me. It is not the case that everybody would be happy learning the guitar or taking life-drawing classes, but it is the case that everyone has something that they would want to spend their time doing. Pretty much anything that starts out as a hobby has the potential to be explored on a much deeper level, if we have the time to dedicate to it.

Political bias can affect how we view this issue. The assumption that the masses have no interest in creativity, research or education, and that they need a paying menial job to structure their lives, can be a common one in people with similar politics to Boles. It is an example of the superiority bias, which is the belief that a freedom you can be trusted with would be dangerous if extended to everyone else.

Currently, when we feel drawn to pursuing a goal or exploring a hobby further, we are usually prevented from doing so by the need to keep paying the rent. It is possible to earn some money doing what you feel drawn to do, such as playing music, opening a specialist shop or developing an app, but it is rare for the amount of money earnt to be enough to live on. For those without a safety net, those dreams have to be abandoned. When actors such as Gary Oldman, Dame Julie Walters or David Morrissey speak out about the lack of working-class actors, this is a large part of the reason why.

One interesting impact of a Basic Income would be to uncouple the link between high wages and high-prestige jobs. Currently, if you are doing hard but vital work such as caring for dementia patients or maintaining sanitation systems you are unlikely to be paid well. With a Basic Income, people currently doing these jobs will have the option of not doing them any more. As a result, salaries for necessary work will have to rise to convince others to undertake them. Slowly, there will be a shift towards linking thankless work that needs doing with work that pays well. Intuitively, this feels like it must be a positive development, but don't be surprised if it is people with highly paid, high-prestige jobs, such as Nick Boles MP, who campaign most vocally against a Basic Income.

In the medieval world, the question that drove us was 'How can I be saved?' The shift to a more material, scientific world changed that question to 'How can I be happy?' Looking at the hard-working, sensible, empathetic post-Millennials, as they turn away from drinking, illicit thrills and the 'hooking-up' culture of casual sex, it seems that they have moved past 'How can I be happy?' as the question that drives them. To my eyes, what seems to be driving them is the question 'How can I be enthused?'

To be enthused is to find something that you are passionately interested by. It is a feeling of intense pleasurable enjoyment and interest, whose mere existence proves that life is self-evidently worthwhile. It is not dependent on material wealth, or the exploitation of the natural world. To be fabulously rich but have no enthusiasm is a sorry life indeed. As the metamodern generation understand unconsciously, it does not matter what you are enthused by, just so long as you are enthused by something. The word itself comes from the Greek *entheos*, meaning 'to be possessed by a god'. To be enthused is also the answer to the questions 'How can I be saved?' and 'How can I be happy?', so perhaps we are becoming wiser as a culture.

Those whose success has come from a background of privilege can be dismissive of enthusiasm. Sir Joshua Reynolds, the eighteenth-century establishment portrait artist and founder of the Royal Academy of Arts, once wrote that 'mere enthusiasm will carry you but a little way'. His book was read by the working-class visionary William Blake, who scrawled in the margins, 'Damn the Fool, Mere Enthusiasm is the All in All!'

The world of paid employment, with its stresses and demands on time, works against our attempts to follow our enthusiasms. This is not the case with Basic Income. It allows people to work fewer hours and gives them the option of turning down work that they do not feel serves any purpose, such as cold-calling telephone sales. More importantly, it allows us to pursue our enthusiasms and see where they take us. These may at first be only of interest to ourselves, but they are likely to be valued by others with similar interests, and this could lead to new businesses and unexpected commercial success. To encourage our enthusiasms, and to avoid blocking them, is to increase the number of people who find their lives meaningful. It is, therefore, the moral choice.

Even though I am not as certain as some that the advance of technology and AI will decimate jobs, I recognise that there is a considerable body of economic argument that supports Basic Income. In April 2018 Mark Carney, the governor of the Bank of England, described the potential benefits of automation. 'On one level the economy can become much more distributed, much smaller scale, much more entrepreneurial, much more bespoke, people can be more creative, people could be more empathetic,' he said. 'And that's actually a pretty exciting or interesting economy, it's a more varied economy, it's a more diverse economy, if you help people get it right.' Basic Income does seem the ideal model for creating that future economy he describes.

But it is the moral question that Boles raised that I think is more important. While Boles's generation may see the moral issue as an argument against Basic Income, those raised in the twenty-first century are more likely to see it as an argument in favour.

8.

Historically, it has needed wars or major disasters before radical changes could pass the splutter test. It took the First World War before votes for rich women and poor men became a reality, and it took the sacrifice of the population during the Second World War before a National Health Service was founded. Both these ideas faced opposition from centrists and, in the case of the NHS, from doctors and the medical profession who found the status quo profitable. Of course, once those changes had been implemented, the centrists found themselves on the wrong side of history. Whereas doctors once felt that a National Health Service would be bad for them financially, now young doctors are inspired to join the profession by the ideals of the NHS itself. Changes like these appear utopian and impossible before they happen. Afterwards, they appear unremarkable.

Basic Income is an idea that appeals to both right-wing and left-wing people, but that does not mean it has broad appeal across the political spectrum. The difficulty is centrists, who fear the radical change such an unproven system would represent. They might not believe in the coming automation that would make such a change desirable, or they might feel confident that technological and cultural changes will not affect them or their livelihood. They might just not like change, and particularly societal change on a large scale.

Will it require a major crisis for radical ideas such as a Basic Income or Half-Earth to become a political reality? Like climate change, some problems are easier if they are tackled sooner rather than later. Implementing a major new economic system would be much easier without having to deal with mass unemployment and environmental collapse at the same time.

One of the reasons why the electorate has not demanded that issues like climate change become a political priority is that they are more concerned with the present than the future. This focus on the present moment is a defining aspect of the postmodern world that most of the present electorate were raised in. While the modernism of the early twentieth century was firmly focused on the future, postmodernism, despite its willingness to raid the past for raw materials, kept its focus on the here and now. Its inability to imagine a future can be seen in the tone of Hollywood movies from the 1980s onwards, the American author Francis Fukuyama's famous declaration that we had reached the 'end of history', and the focus on short-term profit over long-term stability that led to the 2008 global economic collapse.

Now, though, we have moved beyond the postmodern period. Metamodernism is described by academics as 'multi-tensed'. It is all over the shop. It is just as likely to focus on the present as it is the past and the future. A teenager at a party who avoids being 'in the moment' by uploading photos to social media which will be seen by friends in the future, while also becoming the historical representation of the avoided 'now', is a good example of the multi-tensed metamodern world.

All this suggests that the post-Millennial generation will have a different perspective on the future than those of us raised in the twentieth century. At the time of writing the 'climate kids', as twenty-one members of Generation Z have become known, are pursuing legal action against the American government for its failure to prevent climate change. So far, US judges have declined attempts by both the Obama and Trump administrations to have the climate kids' action dismissed. The role of children in climate activism is becoming increasingly visible, such as the 15-year-old Swedish climate activist Greta Thunberg, who stood in front of the December 2018 U.N. plenary session in Katowice, Poland, and told them, 'You are not mature enough to tell it like it is. Even that burden you leave to us children.'

The climate kids are displaying a similar proactive attitude to the survivors of the 2018 mass shooting at Marjory Stoneman Douglas High School in Parkland, Florida. This attitude contrasts with the reactions of older generations who could not imagine a future where things would change, and hence simply resigned themselves to global warming and more mass shootings. If nothing else, it shows that they are a generation who think about the future in a way that the postmodern twentieth-century generations have avoided doing. When over 100 youth activists came together in August 2018 for the inaugural International Congress of Youth Voices, they released a group manifesto which stated that, 'As young activists, we must recognize empathy as a medium that connects us to the world in order to enact positive social change.' Will those raised in the twenty-first century be more likely to contemplate such radical changes as Half-Earth and Basic Income without needing major catastrophes to make these changes possible?

The Dark Mountain people are right when they look at the collapse of biodiversity and climate change and declare that civilisation will collapse if we continue on this path. To those of us raised in the twentieth century, unable to imagine a better future, it is hard to imagine that we will change. But that is a quirk of the age we have lived through. Our vision has been reduced by the dark cloud we were unlucky enough to grow up under.

To the metamodern generation, a future can be imagined, and if it can be imagined it can be built. According to a major global survey funded by the Bill & Melinda Gates Foundation, young people are more optimistic about the future than adults in every country. There are options on the table, regardless of whether or not anyone takes them. Simply discussing them could lead to better ideas, and further options. I don't know what chance our civilisation has in the years ahead, but it is not true to say that that chance is zero.

In 1957, the American psychobiologist Curt Richter performed a cruel experiment in which he left rats to drown in containers half filled with water. He then reran the experiment and briefly rescued the rats at the point when they gave up swimming, before putting them back in the water. These rats went on swimming an amazing 240 times longer than the original rats. Once it is known that there is hope, people seize on this and keep fighting.

9.
MORE THAN INDIVIDUAL

1.

In 2017, the United Arab Emirates announced their intention to build a city on Mars, roughly the size of Chicago, by 2117. The name of this new Martian metropolis will be the City of Wisdom.

This is a hugely ambitious project for a state with no experience of space exploration. Many people in the West were sceptical and openly mocked the UAE's announcement, but I am cautious about underestimating what they might achieve. When the Soviet Union made the first great strides into space, they were a country still recovering from the damage inflicted during the Second World War and they operated under Stalin's unworkable Five Year Plans. The UAE, in contrast, are very rich, and can learn from what other countries have already achieved. In *Red Mars*, Kim Stanley Robinson's epic novel about the settlement of Mars, the planet is originally colonised by Western nations. These settlers are soon joined by people from other countries, who bring the politics and divisions of Earth to this new world. The desert-dwelling Arab people successfully achieve a nomadic, Bedouin-like life there, because they have a particular affinity with the great empty wastes of the red planet.

When the project was announced by Dubai's ruler and the UAE's vice president, Sheikh Mohammed bin Rashid Al Maktoum, a VR simulation of how this future city would look was released to the world. For all my scepticism about large-scale Martian living and the subtle manipulation possible in VR, I couldn't resist it. I downloaded the VR world and was taken on a tour of the future City of Wisdom. The experience felt a lot like a passive videogame, but it did its intended job of making the idea seem more plausible. It's a lot easier to believe in a place when you have been there.

Afterwards, I felt that I needed to balance my VR experience with a more realistic account of what life on Mars might be like. I downloaded a six-part podcast series called *The Habitat*, which told the story of six young people spending a year in a small domed building in Hawaii to simulate what life as part of a Martian crew would be like. The experience of hearing these people talk openly about the difficulties of spending so much time in such close proximity, and the relationships and rivalries that this entailed, was a stark contrast to the UAE's exciting VR experience. After getting to know the six personalities involved, I found I trusted their accounts of their experiences in a way that I could never trust the City of Wisdom tour. Just hearing people talk about what they had been through was more powerful than the technical fireworks of VR. I will remember their descriptions of farting in spacesuits long after I have forgotten the City of Wisdom.

What came across most strongly from listening to *The Habitat* was that NASA's original criteria for selecting astronauts need to change. NASA had previously searched for those high-performing individuals who possess *The Right Stuff*, as the title to Tom Wolfe's 1979 account of the Mercury programme put it. The 'right stuff' referred to the mental and physical characteristics of sharp, reliable, trustworthy astronauts, who could keep a cool head in a crisis and who would always follow orders. What the experiment documented in *The Habitat* showed was that these people were not ideal companions for long-term missions. Those square-jawed heroes could get annoying if you were cooped up with them for months on end. What was needed was people with sufficient emotional intelligence to understand what those around them were going through. This, it turns out, is part of a completely different skill set.

This was one more example of a shift that I keep seeing across our culture. Viewing heroic isolation as desirable has become a dated and delusional world view. The concept of twentieth-century individuals is becoming increasingly inadequate or irrelevant in the complex networked modern world. What is needed now are empathetic people who work well with others and understand themselves as part of something larger.

In David Lynch's 1990 road movie *Wild at Heart*, Nicolas Cage's character, Sailor Ripley, defines himself by his snakeskin jacket. As he explains, 'this here jacket represents a symbol of my individuality and my belief in personal freedom'. The audience viewed Sailor as an admirable and iconic role model when the film was released, but to the eyes of those born in the twenty-first century he is almost unbearably cringeworthy. The jacket supposedly symbolises an isolated, snake-like individual who is beholden to no one, yet this fails to recognise the extent to which Sailor's identity, actions and behaviour are reactions to other characters, in particular those played by Laura Dern, Diane Ladd and Willem Dafoe. The snakeskin jacket, to twenty-first-century eyes, doesn't symbolise Sailor's identity. It represents the opposite. It symbolises his delusional blindness about who he really is.

2.

Wanting to talk to someone about this shift beyond the individual, I decide to pay a visit to the theatre director Daisy Campbell. As I walk over to her house, I reflect on the dangers inherent in such a visit. Daisy has a strange ability to convince people to agree to things. Usually these are things which are bizarre, and which require a serious time commitment. Quite how she manages this has been much debated by people in her circle. The consensus is that she doesn't even know she's doing it. You simply begin a conversation, and this leads to grand schemes and what she calls 'potty capers' being described with glee. Before you know it, the enthusiasm and potential that she wears like a cloak has hypnotised you into thinking that your adoption of some part of the workload is self-evident and preordained.

The innocent audience members who turn up to Daisy's shows are not safe from her influence. A number of people who saw her stage adaptation of Robert Anton Wilson's autobiographical book *Cosmic Trigger*, for example, walked out of the production convinced that they must publish fanzines, create music or put on festivals inspired by what they had just seen. *Welcome to the Dark Ages*, a three-day event in Liverpool in 2017 that she directed, also led to a sizeable percentage of the audience writing books, building websites and producing albums about the events of those days. It's not that Daisy asked anyone to do these things, to be clear. But I think everyone involved would accept that, on some level, she makes them do it.

According to the style watchers at *VICE* magazine, there is a creative movement happening in the UK that, frustratingly for the marketing industry, is almost impossible to co-opt or brand. Daisy and those she has inspired sit firmly at the heart of the theatrical wing of this movement, which the journalist John Doran has named 'New Weird Britain'. This loose moniker covers a diverse sprawl of strange underground events which are linked through certain shared attitudes. These include the blurring of the line between audience and performers, a bewildering level of cosmic ambition on a shoestring budget, and the inability of those present to express exactly what was going on to those who weren't there. As Doran writes, 'They are shows often, but not solely, put on by and largely attended by women. They are shows that many single-minded rock, pop and hip-hop dullards will refer to as pretentious. They are shows that defy a lack of money and other resources by drawing upon deep wells of creative ingenuity – events that are so cost-ineffective, so confounding and so resistant to sane description and promotion that they are impervious to all attempts at commodification. They are shows that cause the participant to stumble back through the permeable membrane wall into reality afterwards, muttering to themselves: "What the actual fuck?"'

I find Daisy in her kitchen, seemingly serene despite the two inexhaustible dogs bounding around her ankles. She thrives on chaos, which is fitting for someone whose middle name is Eris.

Daisy is heavily influenced by the work of the American anarchist author Peter Lamborn Wilson, who writes under the name Hakim Bey. Bey has defined a concept called Immediatism. In his 1994 book of that name, he explores how all experience is mediated and how the connection between people is frequently filtered through some form of intermediate agency. 'Mediation takes place by degrees,' he writes. 'Some experiences (taste, smell, sexual pleasure) are less mediated than others (reading a book, looking through a telescope, listening to a record). Some media, especially "live" arts such as dance, theatre, musical or bardic performances, are less mediated than others, such as TV, CDs, Virtual Reality. Even among the media called "media", some are more and others are less mediated, according to the intensity of imaginative participation they demand. Print & radio demand more of the imagination, film less, TV even less, VR the least of all – so far.'

The more mediated something is, the glossier, more technically refined and superficially impressive it can be, but the further away from genuine human contact it becomes. Thinking back to my reactions to the hi-tech City of Wisdom VR and the simple audio recording of *The Habitat*, I find this idea easy to accept.

It is the least-mediated experiences that mean the most to us, Bey argues, even though these are not what our society promotes or rewards. People feel compelled to create because they feel a need to share or express human, unmediated experiences, yet the more successful and recognised a creative person becomes, the further removed their work becomes from what it was that inspired them.

As Bey sees it, it is vital that we do not lose touch with the spontaneous, playful side of ourselves. To do this involves under-taking projects with no commercial value solely for the amusement of you and your circle of friends. This is the practice that Bey calls Immediatism, which he defines as being immediate and spontan-eous as well as unmediated by technology or organizations. As Bey writes, 'We intend to practice Immediatism in secret, in order to avoid any contamination of mediation. Publicly we'll continue to work in publishing, radio, printing, music, etc., but privately we will create something else, something to be shared freely but never consumed passively, something which can be discussed openly but never understood by the agents of alienation, something with no commercial potential yet valuable beyond price, something occult yet woven completely into the fabric of our everyday lives.'

The basis of Immediatism is the coming together of a group of people who are not your close family or work colleagues, whom you don't have any obligations to or responsibilities for. This coming together is the most difficult part of the process, for the enemy of Immediatism is the busy nature of modern lives. People will always find that they have good reasons not to come together. It is vital that they do, though, because, as Bey writes, 'Of course one must go on "making a living" somehow – but the essential thing is to make a life.'

Once your group come together, a project will appear. This is guaranteed; there is always a project. It will not necessarily be an artistic project in the normal sense of the word. It could be the planning of a fabulous feast, embarking on a road trip or making a quilt. It will be a project of no obvious external value, except for all those involved, who will find their lives enriched by the experience. Ideally, it should not be recorded in any way. The only evidence for its existence should be how people have been affected or changed by it.

'I've been into Hakim Bey since the age of fifteen or something,' Daisy tells me, 'when my dad gave me a copy of his book *TAZ*, which stands for Temporary Autonomous Zone. And it just blew my mind. *TAZ* was my bible. It is mind-altering stuff. I find it takes you down very weird rabbit holes, and I just completely adore his writing. In one of his essays he says, let us all agree that actually one extraordinary night spent in the company of the most extraordinary people, where everything just aligns and ideas are hatched, this joyous coming together, is more vibrant and important than a whole year's worth of politics and proper art, in terms of its getting under your skin and impacting you and giving you fuel to keep going with whatever bizarre mission it is that we're all on.

'I feel like *TAZ* in particular and that book *Immediatism* as well have been trusted companions for a really really long time. It's revolutionary thought, but it's also freeing, and I think very edifying as well, particularly creatively. And I would have been out there, spreading the word of Bey, because too many people don't know his work, but it was only recently that I heard about, you know, the other stuff.'

'The other stuff' is the elephant in the room when you are talking about the work of Hakim Bey. Bey has not been charged or convicted of any offences, but he has written articles that are seen to promote paedophilia, including poems and articles for NAMBLA, the North America Man/Boy Love Association. This has, obviously, tainted the unrelated ideas he wrote about in *TAZ* and *Immediatism*. Those ideas have been influential, with the Burning Man festival and the Reclaim the Streets movement in particular being inspired by his concept of temporary autonomous zones. But you can understand why few people openly celebrate the man.

If you can separate the ideas from their author, then *Immediatism* does give a useful perspective for understanding the emerging twenty-first-century culture. For example, the algorithms that Facebook or Twitter use to decide what you should see are perfect examples of increased mediation and how it alienates us.

Immediatism also highlights opposite examples, in which advances in technology actually make things less mediated. I was reminded of this watching my kids, who were sat at opposite ends of the living room giggling together over Snapchat. The way Snapchat deletes video messages after they have been viewed, leaving no record, is a very Immediatist feature, but it was more the way they were using the technology that I was struck by. They were filming each other, adding zooms and text according to some arcane set of in-jokes that I had no hope of understanding, and sending these looped video clips back and forth between each other. This somehow evolved into a form of conversation that was entirely bewildering but apparently extremely funny. Thinking back to the effort and cost involved in filming, editing and screening video in the 1990s, I doubt I would have believed back then how quickly and cheaply such technology would develop into something so effortless to use. Projects that would have required great expense and large teams can be achieved easily now, with considerably less energy being wasted on planning and organising. The concept of Immediatism, therefore, is a useful lens to judge whether the influence of technology on our culture is bringing us together or keeping us apart.

'I can explain the importance of Immediatism if I use the metaphor of mycelium to describe an artistic culture,' says Daisy. 'Mycelium is like the root system of fungus. It's a network of white fungal threads that runs underground through the soil. The network can be tiny, or it can spread for hundreds of miles and live for thousands of years. If you get sufficient mycelial threads crossing each other, then a mushroom can pop up above the surface.

'Think of the mushroom that comes up above ground as the obvious cultural thing that occurs, which the mainstream culture can see. It might be in the form of a particular artist or a single book or play or a piece of music or a new genre, or whatever it is – whatever cultural phenomenon that the world above can see, that's the mushroom. But the mushroom is not where you should put your focus. That mushroom only occurs because of all of these threads underground created by highly motivated people. They're not motivated by the usual motivations of fame or success or money or whatever. They are people who are following some compulsion and they don't really know where it's leading, but they're moving ahead with it anyway. In that process they're crossing paths with lots of other people who are similarly working away, not really caring if the above world sees it or not, because that's not what it's about.

'In this metaphor, I think of Immediatism as being like a fertiliser. We need to encourage all these threads down there under the soil, and this is the fertiliser with all the right nutrients and goodies. To keep an underground culture healthy, we all need to be constantly inspiring each other by doing things and creating stuff, so that you get that positive feedback loop where people are inspired by people being inspired.

'Immediatism is the fertiliser of the culture because it's the things we just do for each other. The way I see it, the mycelial threads are like the track you leave when you're getting on with your true work. The more they fertilise and cross, the more likely they are to fertilise a mushroom that will actually impact beyond the individual threads all having a lovely time beneath the ground. And that's great, everybody loves a mushroom. But don't be distracted by the mushroom, because it's the mycelium that matters. That's my convoluted metaphor, anyway!

'It can be bloody hard to do this stuff. It's really hard to keep carving your own weird niche when you don't know where it's going, or why. If other people happen to cotton on to your project and come along and you end up with a few bob – oh lovely! But no one's doing it with that expectation in mind at all. In fact, it's very likely to end up costing us. As Bey put it, "It's a picnic but it's not easy."'

3.

Daisy's mycelium metaphor reminds me of a word coined by the musician Brian Eno. Eno has noted that although we habitually focus on a singular individual genius, such as Bob Dylan or John Lennon, that person would not have made the work they are known for if they had not been surrounded by a web of collaborators and inspiration. He proposed the word 'scenius' to describe these inspired webs of connection, and advised that we learn more by looking less at rare geniuses and more at vibrant sceniuses. In some instances, such as the acid-house movement, a music scene can be more fertile and interesting than its most famous faces.

This shift from geniuses to sceniuses, not just in music but in all aspects of human endeavour, neatly sums up the difference between the individualistic twentieth century and the networked twenty-first. In physics, for example, an individual twentieth-century genius like Einstein could expect to radically further science by their singular endeavours. To make such a contribution in the twenty-first century takes huge teams of scientists, such as the men and women who built and operate the CERN Large Hadron Collider near Geneva.

The writers Henry Timms and Jeremy Heimans use the definitions 'old power' and 'new power' to define the change in the balance that comes from no longer seeing the world in terms of hierarchical power. As Timms and Heimans define those terms, 'Old power works like a currency. It is held by the few. Once gained, it is jealously guarded, and the powerful have a substantial store of it to spend. It is closed, inaccessible, and leader-driven. It downloads, and it captures. New power operates differently, like a current. It is made by many. It is open, participatory, and peer-driven. It uploads, and it distributes. Like water or electricity, it's most forceful when it surges. The goal with new power is not to hoard it but to channel it.' An example of this change can be seen with the shift in power that has occurred between Harvey Weinstein and all those who have accused him of abusive behaviour, supported by the wider #MeToo movement.

This change is the result of people becoming connected into digital networks. Given the media's focus on the negative impact of the internet, it is easy to ignore how empowering it has been. As Timms and Heimans note, 'Today, we have the capacity to make films, friends, or money; to spread hope or spread our ideas; to build community or build up movements; to spread misinformation or propagate violence – all on a vastly greater scale and with greater potential impact than we did even a few years ago.'

An interesting take on this necessity of connections comes from the British philosopher and videogame designer Chris Bateman. Bateman is interested in the extent of the commercial networks that we are part of, and how they are so vast we can never see their full scale. 'An oil rig has scuba divers who repair the metal structure when it corrodes with arc-welders entirely unique to their profession,' Bateman writes. 'But who is deep-sighted enough to think of the factories making hyperbaric welding kits, or compressed air-tank regulators, when looking at a car [and the fuel it needs]?'

According to Bateman, recognising that we are enmeshed in a series of vast networks gives us a new perspective on the nature of individuality. In the twentieth century, the pursuit of individual freedom that Sailor Ripley's snakeskin jacket represented was understood in terms of escaping from the bonds of home, religion or local expectations. Leaving the town you grew up in to move to the big city or choosing your own profession rather than adopting your parents' career or business were believed to be ways to increase your individuality. Yet when society is understood as an overlapping web of immense networks, you realise that the opposite is true. Once you have cut the ties that were unique to you, you become adrift in the same mass-culture corporate networks as everybody else. You are using the same technologies, watching the same entertainment, and your life is shaped by the same structures as the rest of the population. Pursuing individuality in this way results in becoming, essentially, the same as everybody else. The background and relationships that were discarded turn out to be the very things that would have made you a distinctive, unique individual. This might seem contradictory to twentieth-century people, but not to the metamodern Generation Z. From this perspective, there is no contradiction between being an individual and being part of a network. The things that makes you unique are your relationships.

Relationships are constantly evolving works-in-progress. They are four-dimensional, because they exist in time as much as in space. The past gives them strength, and the future gives them purpose. In her 2017 book *Radical Happiness,* the Australian academic feminist Lynne Segal argues that the individual pursuit of happiness is a deeply flawed concept because moments of joy and real happiness can only be found with others. As her book's blurb puts it, 'we have lost the art of radical happiness – the art of transformative, collective joy. [Segal] shows that only in the revolutionary potential of coming together is it that we can come to understand the powers of flourishing.' Relationships are precious beyond their role in forming your individuality because, as Segal points out, they are entirely necessary if you are to find life meaningful.

Once the importance of personal connections has been pointed out, it can seem so obvious that it hardly seems worth mentioning. Yet when you go back to twentieth-century discussions about meaning, including such insightful and moving works as Victor E. Frankl's concentration camp writings, you realise that the idea is frequently absent. In the twentieth century, we were so concerned with the ascent of the individual that we failed to notice that it was communal magic which defined us.

'Think of all this as like planning a party,' Daisy suggests. 'You put on the right music, you invite the right people, you lay on the right food, you can do everything right – but you don't know whether or not it's going to work. You don't know if the hoped-for atmosphere will descend or if the party will become a living thing with a mind of its own. It could just as easily not work. Whether or not it does become a party is ultimately an act of grace. You don't know. You hope. You put the things in place, and it happens or it doesn't.'

I think back to AI researchers connecting their webs of nodes in the hope that something as extraordinary as consciousness might spontaneously pop up out of their network. Something like that does seem to happen to actual neurons, if not simulated software ones. Something similar can happen to networks of people too.

'Grace is the right word, I think, because what happens is a similar phenomenon to what you would have found at the beginnings of religion,' she tells me. 'In transpersonal psychology they talk about the ascending path and the descending path as two different approaches to connecting with God or meaning or whatever.

'In the ascending path, the individual is thought of as looking upwards, trying to make a very personal, individual connection with God. This might take the form of pilgrimages or personal quests or sitting on the top of a mountain or hours of prayer – whatever it might be. If you undertake six hours of meditation every day and do this and do that, then you will attain a particular state. That's the theory. But it's a very personal, individual path which is just between you and the universe. You end up thinking that a spiritual or mystical experience is something that happens only inside your own head – that it's a personal thing. That was how we tended to think about spirituality in the twentieth century, and in the whole Christian era really.

'And then there's the descending path. This is about other people, Earth, nature, animals and life in all its forms. It's inclusive, so you can think of it as stretching outwards, rather than looking upwards. It shouldn't really be called the descending path because it makes more sense when you think of it as a horizontal thing. This is actually a much older conception of what religious experience was about, rather than this much more modern idea of just you and the universe. This is where you find grace. It's like a party.

When you get the coming together of people around the campfire or for festivals, you have shared in-jokes, a sense of building a mythology together, a story and shared beliefs and all these things.

'The word "religion" means "to bind together". That's what happens if you keep going to festivals or church or equivalent, or even coming together again and again to produce acts of Immediatism. This is an ancient, ancient phenomenon. People find each other and the fun starts to happen and these moments occur – you've got to be there because it might be one of those nights when grace descends and there's this buzz. It's almost like a big collective falling in love of the whole group. You can just see that's how religions must have begun, back in the pre-literate world.

'There are those who say that that is the true meaning of the Christian cross. The cross is trying to remind us that both paths are required, the ascending path and the expansive, communal, horizontal path. It not an either/or thing. We're all individuals now and, sure, you can have your personal individual path to meaning. But you need the party as well. Everything's better with a party.'

In their 2015 book *The Age of Earthquakes: A Guide to the Extreme Present*, the authors Shumon Basar, Douglas Coupland and Hans Ulrich Obrist write, 'Even though the internet tends to foster an increased sense of individuality, at the same time it's terrific at bringing people together. At the moment, we don't know which will triumph: the individual or the mob. It might be the biggest question of this century.' This question betrays a distinctly twenti-eth-century world view. The idea that the mob and the individual are a mutually exclusive choice seems strange in the metamodern twenty-first century. In the twenty-first century, the answer is not either/or, it is always both and more.

In St Paul's Epistle to the Colossians, he asked his followers to focus exclusively on the ascending path. 'Set your minds on things that are above, not on things that are on earth,' he wrote.

In the twenty-first century on the brink of climate change, that is terrible advice. Focus on the things that are above by all means, but keep an eye on the Earth as well. You can oscillate between both rather than choosing one and ignoring the other. The metamodern generation emerging now intuitively understand that there is no contradiction between the personal, devotional aspect of spirituality, and also the shared grace of community.

4.

I leave Daisy's house, taking away plenty to think about. I am also relieved that she didn't bewitch me into committing my time to some wild scheme or other. It is only later that I remember that, at some point during the conversation, I agreed to accompany her on a visit to the CERN Large Hadron Collider in Geneva for reasons that have yet to reveal themselves. For a visit to Daisy, this counts as getting off lightly.

It is time to return to the beach. It is time to look at the results of the experiment I have been involved in.

It is a warm spring day, even more precious for coming after a long, bitter winter. I had planned to buy a pasty like the one I ate by the sea at the beginning of this book, but the pasty shop has closed and has now become a shop that repairs smartphones. The constant churn of change has continued its relentless progress while I've been thinking about the future.

I go empty-handed to the shore and sit looking at the grey, silent waves. Part of me feels guilty about this because, with something to eat, you can justify time spent looking out to sea as necessary time spent on a daily meal. Simply sitting and doing nothing is thought of as indulgent in our culture, but it is increasingly necessary. In a metamodern world where perspectives swing wildly from one extreme to another, the only still point that you're going to find is within yourself. At a time of turbulence, we all need to become the calm at the heart of the storm. If it needs such indulgent and unproductive behaviour as sitting peacefully by the sea to do this, then go and sit.

Daisy's talk of Immediatism and the necessity of thinking about groups ties together a lot of the threads I have been considering. The key issue here, I think, is trust.

As the physical and digital worlds increasingly blur into each other, and as AI, VR and AR threaten to make the digital world a larger and more overwhelming part of our lives, the question increasingly becomes whether or not we can trust it. We are starting to understand how these things can be used against us, for the benefit of vested interests, at a cost of lowering the quality of our lives. According to the author and Oxford University lecturer Rachel Botsman, we are witnessing 'one of the biggest trust shifts in history: from the monolithic to the individualized. Trust and influence now lie more with "the people" – families, friends, classmates, colleagues, even strangers – than with top-down elites, experts and authorities.'

Living in a world where you constantly need to question the intentions of those supplying the information you are exposed to naturally shifts you away from those you don't know, and towards people that you know and trust. Such a shift is made more profound by the empathetic nature of Generation Z. When your relationships are understood to be a key part of your own identity, it is natural that they will become more important than the influence of faceless companies. Increasingly, the people we would want to bring us the cultural benefits of the digitally enhanced world are those we already know.

People in my generation were baffled when post-Millennials abandoned TV for YouTube, especially when we watched some of the more popular YouTubers for ourselves. Those bumbling amateurs lacked the skills, research, knowledge and charm of professional broadcasters. What we failed to see is what they offered their audience instead. Being of a similar age and having similar attitudes to their viewers, YouTubers were far more relatable than seasoned broadcasters, and the relationships that they build with their audience tends to be honest, open and genuine. To the young, this is far more valuable than professionalism.

When an app requests permission to access all our contacts and text messages even though it should have no need of them, we can choose not to instal it. When social media becomes flooded with waves of hysterical, contradictory narratives and counter-narratives at key political moments, we can recognise who it is that is deliberately muddying the waters of public opinion, and why. Increasingly, we are realising that we can always walk away and not engage. Like the young abandoning Facebook, we still retain the right of veto.

But if we retreat from untrustworthy organisations for the safety of our peers, will we isolate ourselves from what is really happening in the world at large? Will we be imprisoned by the limitations of our own bubble? This was the question that I wrestled with when I first came down to this beach with my pasty. If the anxious, empathetic young were keeping a protective distance from established media, does it follow that they will miss out and remain ignorant of the wider world?

Alternatively, could they be on to something? Could our immediate networks be sufficient to provide all you need for a good, meaningful life? The only way to find out was to conduct an experiment and see for myself.

To do so, I had to limit myself to a social bubble that was small and in no way a balanced representation of the wider world. For the length of this book, I committed to only talking to people I knew well enough to meet for a beer or a coffee on a regular basis. This seemed like a bold start, but it didn't go far enough. To properly test the limitations of a social bubble, I needed to ensure that my test bubble was truly limited. So I took the plunge and further committed myself to only talking to people who I could meet within a short walk from my house. If that isn't a hardcore attempt to limit my bubble, I don't know what is.

Before this, I had written down a wish list of the scientists, companies and experts I wanted to talk to about the future. These were all deeply knowledgeable and respected experts who would have been able to explain the latest developments in emerging fields like robotics, AI or VR. After returning from the beach, I threw that list away. This seemed a foolhardy gamble, particularly considering what I was planning to replace it with. But I would never know for sure unless I went through with the experiment.

Initially, I thought I was making a terrible mistake. I would have liked to have talked to someone about laboratory-grown meat, for example, because it promises to massively lower the carbon footprint of our food, but I did not know anyone who was involved in that field. Frustratingly, David Bramwell did know someone, and he offered to put us in touch, but that would have broken the rules of my experiment. Occasionally, a moment of good fortune would come to my aid. I had been thinking about what Michelle Olley had told me about VR, and I was frustrated that I couldn't include her thoughts because she lives in London. Then I randomly bumped into her walking down Trafalgar Street in Brighton, which meant she could be included.

Very quickly, however, I found that talking to trusted peers helped me understand all sorts of issues in a way that interviews with experts didn't. It's amazing how well informed regular people are about things that genuinely interest them. Because political discussion is distorted by partisan spin, and news media are designed to exploit our psychological biases, it is easy to assume that the information age is nothing but fake news, ignorance and propaganda. But outside of news and politics, in the world of technology, culture and science, the information age is actually doing a pretty good job. The insights of experts are shared in books, interviews and talks and quickly get disseminated through our information networks to those who are interested, where they are debated and assessed. The internet has meant that people can now be surprisingly well informed about their particular areas of interest, if they choose to be so.

What I learnt from my peers was qualitatively different from what I would have learnt from experts. Leading experts share information in a professional manner that reflects well on the company, university or media outlet that pays their wages. Their words are tempered by the fact that their reputations are on the line. That isn't the case when you talk to your friends. There is an openness to the conversation that comes from an existing level of trust. They do not have an official line that they are duty-bound to repeat. They are free to speak more personally and subjectively when they do not have the same level of skin in the game.

Being a few steps removed from the cutting edge, enthusiasts talk about technology as it filters down to them. They have a pragmatic understanding of what it is really like that differs from the idealised expectations of research laboratories. It is at this level that the first divergences away from the arrow-flight projections start to reveal themselves.

There is another difference which feels important to mention, although it is not an easy thing to explain. It wasn't just the content of the information gained from friends that was different. The act of learning from them was qualitatively different also. The connections between people are mysterious things that are a lot richer than we usually give them credit for. I do not know the word to describe what is found in the gap between people, but I understand why Daisy spoke in terms of grace.

I recall something that the late Beat writer Brian Barritt once told me. The question of life isn't really that hard, he said. The only real difficulty is finding interesting conversation. This echoed something that Carl Jung once wrote, which is that 'Loneliness does not come from being alone, but from being unable to communicate the things that seem important.' According to Rebecca Adams at the University of North Carolina, the factors which are crucial to making friends are 'proximity; repeated, unplanned interactions; and a setting that encourages people to let their guard down and confide in each other'. This used to mean that most of our friendships were forged in school or college, and that making friends became harder once you left education. The internet has put us in a position where the factors Professor Adams highlights have become more commonplace in adult life. For those who have an enthusiasm to pursue, the modern world has made making friends while middle-aged entirely possible.

It is easier to find people you click with these days, but it is still easier not to. With fast internet and endless box sets on Netflix, we now have the option of avoiding others and still never being bored. For a socially anxious generation, the temptation to withdraw can be strong. That many will choose to do so is heartbreaking. The extreme end of this phenomenon is a significant social problem in Japan, where adolescents and adults who withdraw from all social interaction are known as the *hikikomori*. In 2010, the Japanese government estimated that there were 700,000 *hikikomori*, most of whom who are living in a room at their parents' home, with an average age in the early thirties.

I am by nature an introvert and can see the appeal of hiding away. Extroverts are energised by social contact, but introverts are drained by it. I would find it easier to hide away like a hermit, but knowing how important others are to my sense of self stops me from taking this option. Instead, I make sure I know my limits, and I make sure I don't exhaust myself. I balance social activity with the periods of quiet that my particular personality type needs. I know I prefer the company of small groups of people rather than large ones, so my social life is structured around meeting individuals for a drink rather than getting everybody round for a party. In the metamodern world you naturally grasp the limits of things and know when they are not appropriate. But you also grasp the strength of things too. This applies to personal relationships as much as anything else.

5.

An important factor in my experiment was where I live. Brighton, on the south coast of England, is not a typical town. In April 2018, the international relocation website MoveHub analysed 446 cities across 20 countries in an effort to discover which was the most hipster city in the world. The results were calculated by analysing the number of vegan eateries, coffee shops, tattoo studios, vintage boutiques, and record stores for every 100,000 residents. As is the way in our dataist society, five data points are assumed to be sufficient to define the infinitely complex evolving entities that are our cities. While we know those data points are fundamentally meaningless and arbitrary, they nevertheless still create an annoyingly plausible analysis of the city's character. The result was incredibly close, but Brighton was declared the most hipster city in the world, just beating the second-placed Portland, Oregon.

Brighton is seen as an alternative and radical place by some and an immoral hellhole by others. It has been, for many years, the gay capital of the UK, and it is the only place in the country to have elected a Green MP, the remarkable Caroline Lucas. Brighton boasts Britain's first naturist beach, and it was here that the first adult film in the world was shot, *Victorian Lady in Her Boudoir* (1896). When an evangelical priest declared Brighton to be the 'most godless city in Britain' in 2009, a poll in the local paper showed that 93 per cent of the population took this as a compliment. A poll in 2008 declared it to be the 'Happiest place in the UK'.

I love Brighton and find it to be an incredibly welcoming city, but it is not for everyone. The columnist Keith Waterhouse memorably described it as a town that looked like it was helping the police with their enquiries. An editorial in the *Sun* newspaper once called it a 'Town of Shame' and said that 'Brighton has become a nasty town of drugs, gays, AIDS and drunks, with a left-dominated council whose mayoress once refused to curtsy for Royalty.' Yet it would be a mistake to view the place as a left-wing hotbed. Hove, the western part of the city of Brighton and Hove, is more traditionally Conservative, while the city overall is extremely entrepreneurial and lacks large, working-class industries. Brighton is a place of independent traders, expensive property and self-employed freelancers, and a 2018 study named it the best city in the UK to start a small business in. The large amount of digital start-up companies here has led to it being called 'Silicon Beach'. This mix of entrepreneurism and radical inclusive politics makes it a very metamodern place. It comfortably embraces the contradictions of both extremes rather than seeking a compromise, or focusing on a single, consistent aspect.

There is a significant crossover between the values and interests of Generation Z and the culture of Brighton, which has been invaluable for my experiment. Brighton is very accepting of people's desire to express themselves, and so the people I meet all tend to be pursuing their own enthusiasms. In some places I have lived, creative expression is seen as attention-seeking, deeply embarrassing, or a sign that someone has got up themselves and needs bringing down a peg or two. In Brighton, it is entirely normal. The town feels a little like a Petri dish, where the values of the emerging metamodern generation could be studied and tested. This is a place where you can see how emerging attitudes work in a real-world scenario.

It is here, then, that you find the merry crew who are part of the network of relationships that make me who I am. They are the cast of my own personal sitcom. They are people outside of family and work colleagues who I have no responsibility for, but who I choose to be trapped with regardless. If life in the twenty-first century is to be an Immediatist metamodern sitcom, then here is a prototype I cobbled together.

We can clearly see the flaws in such a limited approach. The question is, are the strengths of it sufficient to make it worthwhile regardless? Were the insights of my peers sufficient to produce a plausible vision of the future, or did they prove to be too ignorant and limiting? Was it sufficiently rewarding and interesting? Is what I have learnt of any value, or is my bubble too much of an outlier, and too full of oddballs, to tell us anything about the culture that is emerging?

Here's what I learnt: I have learnt that people change. Gene Roddenberry was right when he insisted that twenty-third-century astronauts wouldn't necessarily smoke cigarettes on spaceships. It is no use assuming that your own biases and prejudices will be common in the years ahead. We will all look as blind, naive and funny to future people as people from the past look to us. Culture is constantly evolving and being born just a few years earlier or later is enough to give people noticeably different values. We can trust that the culture of any given moment will be a logical reaction to the world at that time. In the words of the cognitive historian Jeremy Lent, 'culture shapes values, and those values shape history'.

I have learnt not to look at science fiction as a road map of the coming future. As the truism goes, science fiction is always about the concerns of the present. Applying arrow-flight projections to modern technology can create some great drama, but it shouldn't be confused with prophecy.

I have learnt that our future is on Earth, not out among the stars. It is fair to argue that perhaps in 5,000 years' time our understanding of science will be so far ahead of what it is now that we will be able to colonise alien worlds. But this assumes that our civilisation will be alive and well at that point. If we don't learn to live sustainably within our biosphere, and adapt to the climate changes that it is too late to prevent, this will not be the case. For the rest of the twenty-first century, then, our challenge is life on Earth, not a 'plan B' of life in space.

I have learnt that, for all we can expect huge advances in technology, our machines will not take it upon themselves to rule us. If these tools are used to oppress us, it will be because another human is using them to do so. Technology has a strange ability to reveal what it is that makes humanity unique. This is usually positive, but not always.

I have learnt that there are great imbalances in the ecological and economic worlds, which will have to be tackled. This will be far from easy, but it is not impossible, and there are potential ways forward on the table. While the powerful individualism of Baby Boomers, Gen Xers and Millennials have prevented them from taking the necessary steps, I do not think that those born in the twenty-first century will be quite so happy to hide from reality.

I have learnt that we should not be so quick to dismiss this 'snowflake' generation. They understand that we are not isolated individuals, and that our relationships are central to who we are, how we think and what we do. This realisation seems crushingly obvious once it has been internalised, but it has been notably absent in twentieth-century thinking. The shift from cooler-than-thou isolated individuals to involved, enthused connected people brings meaning into our visions of tomorrow. This immediately erodes many of the assumptions implicit in the twentieth-century nihilistic vision of the future. The idea that people will give up and not actively attempt to solve our problems looks increasingly unlikely in a population that has something to live for. This shouldn't be too much of a surprise. It is, after all, how we have predominantly acted throughout history.

I have learnt that we do have a future. Not an ideal future, obviously, as we have left it too late to comfortably tackle climate change and the collapse of biodiversity. But I no longer share our culture's belief that everything will come to an end.

From our films, commentators and social media, I had thought that a collapse was coming, and the only question was when. Now, I think that I happened to be born just as a wave of nihilism was washing over our culture. That wave blinded my generation and prevented us from seeing what was ahead. This wave, I think, is starting to recede. To give one small but telling example, the fact that no film or TV company had ever tried to adapt Iain M. Banks's hugely influential *Culture* series of novels always felt significant. Banks's future Utopia was too much at odds with the circumambient mythos to risk spending millions of dollars filming them. But in February 2018 Amazon's CEO, Jeff Bezos, announced that Amazon Studios would adapt the books, claiming that he himself was a massive fan. This is just a small incident, of course, but if you look at Generation Z culture you will find many such small incidents. It does seem that a new circumambient mythos is forming.

There are winners and losers in every era. Some people were

ideally suited to life in the Renaissance, whereas others thrived during the Industrial Revolution. The coming world will be no different. There will never be a complete Utopia or dystopia. There are always only changing, shifting situations that benefit some more than others. In past eras, it was a disadvantage to be weak, or uneducated, or workshy. The years to come, I suspect, will not suit those who are inclined to be loners, critics or passive consumers. The attributes that will be most useful are creativity, imagination and playfulness, combined as always with fidelity and tenacity. We will need to be self-motivated and curious, and we will need to be able to forge relationships which survive long periods of time. We will need the ability to be interested – it doesn't matter what in, just interested in *something*.

I still can't predict the surprising events, unexpected inventions and political disasters that the years to come have in store, but I no longer feel that a default reaction of despair is appropriate. That, ultimately, is what I have learnt from this experiment. It is too early to say how accurate my attempt to step into the mindset of the coming years has been, but even if it has not left me well informed, it has certainly left me enthused.

6.

Whose responsibility is it to make a future worth living in?

The experience of what it is to be human has expanded over the centuries and become richer and more expansive. For most of history, the job of expanding the human soul – for want of a more technical description – has fallen to the rich and the elite. It was these privileged people who explored beauty, luxury and sensuality, and who commissioned and funded the architects, painters and musicians who moved art forward. Since the First World War, however, the rich and privileged have almost entirely abandoned this work. Instead, for the remainder of the twentieth century, the job of advancing what it meant to be human fell to the avant-garde and the counterculture. Because the twentieth century was so focused on individualism, we have tended to credit individual geniuses with these advances, even where it was sceniuses that deserved the credit.

Since the emergence of the internet in the mid-1990s, however, the vibrancy and the originality of the counterculture and the avant-garde have dissipated. Now, the work of evolving and improving the human experience falls to everyone. No one is going to produce meaning for us. We are all responsible for our own individual part of the culture. To live a life worth living, it is necessary to make it worth living. To twentieth-century people this might seem like a lot of responsibility for a single individual, but we are more than just ourselves alone. Our individuality is defined by our relationships. It is here that meaning and culture grow.

Our relationships are part of us. They are needed to develop our skills, and we could not produce our achievements without them. Connections are a map of our reputation and they make us unique. In a global village where we are all in competition with the whole world, that distinctiveness is our unique selling point. This alone is a reason to value relationships more highly than objects or status symbols. But their true worth is perhaps more important than this.

Biodiversity flourished at Knepp once the top-down attempts to manage the land ended. Instead, every individual aspect of the ecosystem was trusted to act in response to all that it was connected to. Impediments to this process were removed, and the ecosystem flourished. It is now stronger and more vibrant than any top-down management plan could ever dream of. Culture is the same. Stop trying to enforce it from above, remove what stops people from expressing themselves and allow the connections between people to deepen. The culture that springs back will amaze us all with its richness and meaning.

If you nurture your relationships well, what results will be that most precious of treasures: a fulfilling life. The rules for how to do this are no secret. They basically boil down to a single commandment: try not to be a dick. If you treat people with respect and concern, and make sure you offer up as much as you receive, then you won't go far wrong. Of course, we are all dicks sometimes. We are all fallible and bewildered. No one is expecting perfection. It would be sensible, though, to try to not make a habit of it. There does not seem to be a lot of forgiveness around these days.

Having something to live for forces us to face the great environmental concerns of our age. Once we've accepted that fleeing the planet and living in outer space is not an option, and that our future is here as part of the Earth's biosphere, that puts us in the unenviable position of having to somehow fix the imbalances that have been building up. We should have started this years ago, but if we trust Davies's corollary there is still hope: 'Things that can't go on forever, go on much longer than you think.'

Is this possible if the great chain of being remains at the heart of our circumambient mythos? Sadly, I'm not certain it is. To replace it, we will have to see ourselves as the equal of nature instead of its master. Mankind and nature will have to make allowances for each other. Both need to make sacrifices for the long-term good of the relationship, particularly as the changing climate forces people to adapt. But, like in a good marriage, knowing that you are building something strong that will sustain across long periods of time makes compromises easy to live with.

This shift of focus towards networks also brings with it the potential for tribalism. Wherever there is competition for scarce resources, there you will find conflict between groups. But in this new world view the resources of primary importance are meaning, enthusiasm and relationships. These are not limited resources. There is an infinite supply of these things, should circumstances be arranged in such a way that we can dedicate our time to unearthing them.

Those circumstances include access to food, shelter and tools. Unlike meaning, enthusiasm or relationships, these are currently in limited supply. It would benefit everyone if we all had access to life's necessities, because it would cause the clashes between tribes to fall away.

It is possible to arrange society in such a way that everyone would be fed and housed, if we all wanted to do so. If a system such as Basic Income does become a reality, and the energies currently spent on survival become available for enthusiasms, relationships and the generation of meaning, then the dissolution of the great chain of being becomes easier to imagine. Gene Roddenberry thought that homelessness and poverty were not eternal laws of nature but problems that we can overcome, just like slavery or absolute monarchy. If he was right, then support for a Half-Earth initiative and a healthier relationship with a natural world starts to look plausible. These seem to be ideas worth exploring, improving and campaigning for. If nothing else, they will lead to more interesting conversations than just giving up, complaining and waiting for doomsday.

7.

And what of the children being born now? They will, in due course, come to be defined as the generation that follows Generation Z. These children, the post-post-Millennials, will have their own fresh take on the world. They will be shaped by the culture their elders have left them. Quite what their priorities and fears will be is entirely unknown, but if we look back at preceding generations there is a pattern which might give us a clue.

Each generation tries to move away from the failings it sees in the older generation, while unconsciously adopting their successes. For example, Baby Boomers unleashed a wave of optimistic creative individualism, but they could be extraordinarily naive about how the world works. When Generation X arrived, they accepted and continued the Boomers' focus on individuality, but they defined themselves as different by avoiding their naivety. When Millennials arrived, they liked the individualism and avoidance of naivety, but they found the ironic preceding generation to be too nihilistic and cynical. Millennials reacted by insisting there was meaning in the world, even if it was only to be found in things that were personally meaningful rather than a great universal absolute truth.

Generation Z, in turn, see Millennials as too self-focused. They understand that while remaining childlike is important and appealing, it needs to be countered with realism. As a result, Generation Z work hard, because they do not believe they will be handed success and recognition just for turning up. They are trying to return to such neglected concerns as financial security and emotional stability. Having grown up during the aftermath of the global credit crunch, at a time when there were no positive stories about the future in their culture, it is perhaps not surprising that they have record rates of anxiety and other mental-health issues.

Given all this, how might the children being born now come to view Generation Z? If history works as a guide, they would unconsciously absorb what is good about Generation Z, which is their marriage of individualism with network thinking that has produced this great surge in empathy. But it also seems likely that they will view Generation Z as oversensitive. Is it possible that the children being born now will react against this and become more stoical and thick-skinned than their immediate elders?

If this is the case, then this could be a remarkable generation indeed. We might potentially be witnessing the birth of a twenty-first-century equivalent to the twentieth century's Greatest Generation, who were born between 1910 and 1924. The Greatest Generation grew up during the Great Depression and went on to fight the Second World War and create the welfare state. They had not been dealt the best of hands, in other words, but they rose to the challenge remarkably. Would implementing ideas such as Basic Income or Half-Earth seem quite so implausible to a generation like that?

When we consider the economic and environmental problems that we will face in the future, it is easy to assume that the people involved will think the same way that twentieth-century-raised people do. When we predict their actions and priorities, we think what we would do in their circumstances. But it's not enough just to place ourselves in the shoes of these future people. We have to place ourselves in their minds as well.

When those who lived through the Second World War tried to imagine the future, they never imagined that it would contain gay marriage or trans rights. When our values and prejudices change, they can change rapidly and unexpectedly. The combination of empathy for others and a zero-tolerance policy for those who cause harm that is emerging now will greatly affect the politics of the later twenty-first century, once Generation Z have replaced Baby Boomers in the voting booth.

It is easy to overstress the size of this generational shift. As noted earlier, those who study demographic attitudes focus on the changes that occur between generations rather than the similarities. It is certainly not the case that the future will be entirely populated by empathic justice warriors. Self-interest, inequality and a dislike of change are not going to disappear. People will still get grumpier, more cynical and warier of change as they get older. Cultures that see individualism as a major part of their identity will struggle with the coming changes. There will be fierce arguments and ideological struggles, just as there have always been. But the values and network-thinking that our children are demonstrating will be a factor in those arguments. It doesn't need a huge shift to tip the balance of arguments from the old approach to politics to the new.

At the moment, after the rise of nationalist populism which led to Brexit and Trump, and which is heavily skewed towards older generations, we are witnessing the twentieth-century world view lashing out and trying to save itself from coming change. This is the last stand of the individualistic, fundamentalist, single-vision philosophy, which calls for walls and isolation like Canute ordering back the waves. The young are watching all this play out. They are not seeing anything that appeals. They are certainly not seeing anything that works.

Sometimes, a virus has to run its course before you can be cured. Some ideologies need to reach their failed, absurd ends before we can get them out of our system. This is how we create the antibodies that will protect us from that virus in the future. That's what's happening now. It's not fun, but it's probably necessary. It will be worth it in the end, in the future that we are all building together, whether we deny it or not.

AFTERWORD TO THE PAPERBACK EDITION

I write this afterword a year and a half after finishing the book, while the world is on lockdown due to the coronavirus pandemic.

Even before the virus hit, the generational and cultural changes discussed had been occurring far more quickly than I expected. You can find countless examples of this change in values. For example, the plot of the bestselling videogame at Christmas 2019, Hideo Kojima's *Death Stranding*, is not concerned with an individual shooting his or her enemies. Instead, it tasks the player with rebuilding connections and establishing communities in a fractured world. When the zombie comic *The Walking Dead* finished its sixteen-year run in July 2019, its story had gone from the despair of isolated and untrusting loners attempting to survive to communities coming together and building a world better than it was before the zombie outbreak. The four shortlisted artists for the 2019 Turner Prize announced they had become a collective, negating the possibility of an outright winner. In cinema, Todd Phillips's billion-dollar hit movie *Joker* triggered a flood of media hot takes and think pieces debating whether it was deeply responsible or highly irresponsible. Being an extremely metamodern film, it was of course both. On TV, the *Star Trek* franchise has started looking to its fictional future again. The importance of this is largely symbolic, but it is welcome nonetheless.

Generation Z's worldview has impacted politics far earlier than I expected. When I mentioned Greta Thunberg at the end of Chapter 8, I did so assuming that readers would not know who she was. Having since been nominated as a candidate for the Nobel Peace Prize and become the youngest individual to be named *Time* magazine's Person of the Year, she is now one of the most famous people on the planet. Thunberg's profile is a consequence of the change in her generation's thinking. Generation Z do not ask, 'What can *I* do about this?', they instead ask, 'What can *we* do about this?' These are superficially very similar questions, but they produce very different answers.

This rapid rate of change has generated an enormous backlash. Thunberg in particular has been on the receiving end of an extraordinary amount of abuse. In retaliation, the young began responding to criticism with the phrase 'OK Boomer'. The website UrbanDictionary.com defines the meaning of this phrase as: 'When a Baby Boomer says some dumb shit and you can't even begin to explain why he's wrong because that would be deconstructing decades of misinformation and ignorance so you just brush it off and say OK.'

There is nothing new about young people being dismissive of older people, of course. Baby Boomers themselves were masters of it. In the 1960s counterculture, the phrase 'Never trust anyone over thirty' was common, and The Who sang about how they hoped they would die before they got old in 'My Generation'. Bob Dylan's 'Ballad of a Thin Man' actively taunted the older generation who just didn't 'get it'. He sang about how something was happening but they didn't know what it was, before adding the sneering taunt, 'Do you, Mr Jones?'

There were indeed many 'Mr Jones's who did not understand the generational change that was occurring. They had grown up in an era that was hierarchical rather than individualistic, and they could not grasp the young's subconscious individualism. This individualism was ingrained in the psychology of the younger generation, so it wasn't something they were going to 'grow out of', as many of their elders assumed.

Those same Baby Boomers are now trying to deal with a similar generational change that has occurred in twenty-first-century youth, because we all become Mr Jones eventually. 'OK Boomer' might seem relatively harmless in this context because there is no insult used, just the accepted name for the post-war generation, yet it has proved to be a surprisingly effective and wounding taunt. By labelling an older person as a typical member of a group – the Baby Boomer generation – the young are denying them their cherished individuality. The same older people, in contrast, use an individualistic word as their preferred insult for the young, 'snowflake', to label them as single, isolated things who are weak and helpless. These contrasting approaches to insulting generations cuts to the heart of our current divide, in which people raised in the twentieth century generally see themselves in individualist terms while those raised in the twenty-first century see themselves as part of a network of relationships, without which their individuality is meaningless. They understand that, yes, they may be snowflakes, but at the same time they are part of an avalanche.

It might seem a minor, trivial issue, but a small change to such a deeply buried foundational aspect of our psyche has huge implications. On the face of it, there is no reason why an opinion on Brexit, which is a question of political sovereignty, should negatively correlate with concern about climate change, which is a question of understanding science. In practice, however, it is extremely rare to find someone who admires both Nigel Farage and Greta Thunberg, and it is very easy to find hundreds of people who approve of one but not the other. The demographics of this split, it turns out, are a close match with the demographics of the individualist/networked divide. Worldviews that differ on such a fundamental level are bound to clash, and a divide like this does not lend itself to dialogue and compromise. The dissonance frequently emerges as anger.

Previously, the UK was traditionally divided politically in terms of class, but now age is the best signifier of how a person will vote. Among eighteen- to twenty-four-year-olds at the 2019 UK general election, 19 per cent voted Conservative and 57 per cent voted Labour. These figures are almost the mirror opposite of those aged over sixty-five, of whom 62 per cent voted Conservative and 19 per cent voted Labour. Although Millennials are less likely to vote compared to older generations, the teenage Generation Z cannot wait to get in the voting booth. Around 750,000 members of Generation Z become eligible to vote each year, while 500,000 mostly older people die of natural causes. It will not be long before Generation Z tip the scales in the political make-up of the country.

In climate terms, however, this delay is critical. Scientists are adamant that we need to decarbonise our economy now if the worst climate scenarios are to be avoided, but the older generation, who are courted by political parties because they can be relied on to vote, are more likely to resist the changes needed. At the start of 2020, the generational clash looked like it would only intensify, and get a lot more unpleasant than the phrase 'OK Boomer'. Being based on a fundamental change in worldview, understanding the interconnected nature of the economy and the environment was not something Generation Z were going to 'grow out of'. From their perspective, the old were actively damaging their future. It's normal for kids to be rude about their parents, but for them to have such a negative view of their grandparents was something new.

Then the coronavirus pandemic arrived. At the time of writing, it is not known how long the lockdown will go on for, but very few people expect the world to be the same afterwards as it was before. The enforced lockdown has been like a global monastic retreat, and people will emerge from their isolation with their values and goals changed. Our views on which jobs are important has been radically altered, for example, and it is clear that it is not the jobs that pay the most that are now the most valued. The pandemic has highlighted the importance of our relationships, because the connections between people are the same vectors along which the virus travels, so our attention has been focused on those connections like never before. This book has argued that in the twenty-first century we will focus more on our networks of relationships than we did in the twentieth, and that argument seems a lot less controversial now.

One result of people being forced to isolate is that we realise how much we value each other. This has helped soothe the growing intergenerational conflict. Because the virus is more dangerous in older age groups, the young have put their lives on hold, and are experiencing great financial hardship, in order to protect the older generation. They recognise that, for all their frustrations about the older generation's politics, they still value them as people. Of course, the young also expect to be valued and considered in return, and that includes their concerns about the climate. Some older wealthy people used to say that they absolutely needed to fly several times a year and there was no possible way they could change their behaviour. This is not an argument that they will be able to get away with after this.

It may be that our inertia in dealing with the issue of climate change will prove to be another victim of the virus. It is not simply that we have been shown a different world, with cleaner air and the return of wildlife to places where human activity previously kept them away. It is that the global economy faces record unemployment and a prolonged economic depression, and there is a notable absence of economists arguing that austerity and inflating asset values will solve these problems. Instead, governments are being urged to take advantage of record low interest rates and to spend huge sums of money in order to create jobs and stimulate the economy. The question is, what will they spend that money on? Establishment voices, ranging from current and former central bankers to Prince Charles, are arguing that this money needs to be spent as part of a Green New Deal to undertake the previously resisted major infrastructure changes needed to decarbonise our economy. They argue that green stimulus would create many jobs and utilise existing technology, and that there are many 'shovel-ready' projects that can begin immediately. Almost miraculously, the necessary political action needed to tackle climate change has suddenly become possible.

There were many ideas for a better future which people dismissed as unrealistic before the pandemic, but which now seem necessary and sensible. The situation can be likened to how all the great modernist works appeared after the First World War, even though all the key ideas behind those works were developed before it. The war removed inertia and tradition, and suddenly all these new ideas and values were free to run wild. In a similar way, many ideas suggested in this book are suddenly gaining support from institutions that would previously have resisted them.

An important example is basic income, which looks like it will be implemented in Spain. The Spanish scheme is not a universal basic income in its purest form, in that it is intended to be targeted rather than given to the entire population, but it is still a major step forward. What has helped the argument for basic income is that we now have a much better understanding of which jobs are important, and we can see how the old system has failed to value or financially reward what matters. It will be interesting to see how Spain gets on, and who will try the idea next.

Despite the pandemic, the rate of societal change is still increasing. Technology continues to play a large part in this, and it has continued its haphazard, drunken advance since this book was written. Elon Musk's *Crew Dragon* spacecraft has now successfully ferried astronauts up to the International Space Station. The VR game *Half-Life: Alyx* has proved to be a major step forward for virtual reality. The wild claims being made about AI have started to die down, as we become more aware of the limitations of the technology, and the extent to which the rate of automation will be affected by high unemployment is currently uncertain.

But technology is just one element in the changes we're living through. Society is also evolving, and so are our values, beliefs and ideas. The combined pressure from the pandemic and generational change has thrown off our previous certainties and given us a rush of options. The future is up for grabs now, and perhaps the fact that we have lacked an idea of the future will turn out to be a blessing. We are not limited by an out-of-date road map. We have the freedom to react, experiment and make it up as we go along.

Over at the Knepp Estate, the rewilded landscape continues to grow in biodiversity. In May 2020 white storks hatched six stork chicks there, in two stork nests. This was the first time that wild white storks are known to have successfully hatched eggs in this landscape since 1416. For those fond of signs and symbols, the return of hatching storks after six centuries was surely as good a sign as you could wish for. A new world is being born, regardless of those who deny it, and it falls to those alive in the twenty-first century to both witness it, and to shape it.

ILLUSTRATIONS

By Melinda Gebbie

NOTES AND SOURCES

INTRODUCTION

1.

The Robert Zemeckis quote is taken from the 2010 documentary *Tales from the Future: Time Flies*, on the *Back to the Future Part II* Blu-Ray. The quote from Adam Sternbergh appears on p. 197 of *Radical Happiness* by Lynne Segal. For details on how close we came to ATM machines being switched off see Andrew Grice's interview with the former Chancellor Alistair Darling in the 19 March 2011 issue of the *Independent*, 'Alistair Darling: We were two hours from the cashpoints running dry'. The 19 June 2018 article by Patrick Collinson in the *Guardian*, 'Visa admits 5m payments failed over a broken switch', gives an account of the failure of VISA cards.

2.

The quote from Brad Bird comes from Steven Zeitchik's 15 May 2015 *LA Times* article, '"Tomorrowland" director Brad Bird keeps looking for the bright side'. The *New York Times* review of *Tomorrowland* was written by A. O. Scott on 21 May 2015, and the *Philadelphia Inquirer* review, also from 21 May 2015, was by Steven Rea.

3.

The use of Yuval Noah Harari's term 'the cognitive revolution' should not be confused with the advances in psychology, linguistics, anthropology and related fields in the mid-twentieth century that are also known as the cognitive revolution. For more details about the Stadel Lion Man, see the British Museum blog at blog.britishmuseum.org/the-lion-man-an-ice-age-masterpiece. For more about Earth Overshoot Day see the website www.overshootday.org. The Robert Nisbet quote is from his book *History of the Idea of Progress* (1980), although I took it from p. 40 of Steven Pinker's *Enlightenment Now*. The Christopher Booker quote about tragedy is taken from p. 155 of his book *The Seven Basic Plots*, while his description of comedy is taken from p. 150.

4.

Barack Obama's 2016 quote is taken from p. 37 of Steven Pinker's *Enlightenment Now*.

1. ON BEING REPLACED

1.

The 2013 University of Oxford report *The Future of Employment* is online at https://www.oxfordmartin.ox.ac.uk/downloads/academic/The_Future_of_Employment.pdf. For a report on Andy Haldane's comments, see Larry Elliott's 12 November 2015 article in the *Guardian*, 'Robots threaten 15m UK jobs, says Bank of England's chief economist'. The 2017 PricewaterhouseCooper (PwC) report can be found at www.pwc.co.uk/economic-services/ukeo/pwcukeo-section-4-automation-march-2017-v2.pdf. The website of Flippy, the burger-flipping robot, is misorobotics.com/flippy. For details of the AI that performed better than lawyers, see Monica Chin's 26 February 2018 article on Mashable.com, 'An AI just beat top lawyers at their own game'.

For a report on Anchor's survey into the elderly's attitudes to automated tills, see Sean Coughlan's 21 November 2017 BBC News report 'Automated checkouts "miserable" for elderly shoppers'. The *Vanity Fair* article quoted is 'Why Hollywood as we know it is already over', by Nick Bilton, dated 29 January 2017, and the Lee Child quotation is taken from Andy Martin's 2015 book *Reacher Said Nothing: Lee Child and the Making of Make Me*. For more on the financial markets' reaction to claims of an AI and robotic revolution, see Chris Dillow's 29 December 2017 article in *Investors Chronicle*, 'The robot non-revolution'. The July 2018 PwC report which claims that such technology will produce a net gain of 200,000 jobs is discussed in a p. 5 story in the 21 July 2018 issue of *New Scientist*, 'Robots won't be taking your job'.

3.

The quote 'Neurons that fire together, wire together', also known as Hebb's law, is a catch-phrase or summation of ideas explained in more detail in Donald Hebb's 1949 book *The Organization of Behavior: A Neuropsychological Theory*. The conversation between Larry Page and Kevin Kelly is recounted on p. 36 of Kelly's book *The Inevitable*. As noted above, the *Vanity Fair* article quoted is 'Why Hollywood as we know it is already over', by Nick Bilton, dated 29 January 2017.

4.

If you wish to listen to algo-generated black metal, you'll find the *Coditany of Timeness* album streaming at dadabots.bandcamp.com/album/coditany-of-timeness. If you would prefer to read *Harry Potter and the Portrait of What Looked Like a Large Pile of Ash*, you'll find it at botnik.org/content/harry-potter.html.

2. ARROW-FLIGHT PROJECTIONS

1.

The 1 August 2017 'ROBO STOP: Facebook shuts off robots after they chat in secret code' article in the *Sun* was by James Beal and Andy Jehring.

2.

For more on Google Duplex, see the 8 May 2018 entry at ai.googleblog.com, 'Google Duplex: An AI system for accomplishing real-world tasks over the phone', by Yaniv Leviathan and Yossi Matias. The quotes from Microsoft's *The Future Computed* 2018 report are taken from pp. 37 and 43. For an overview of Chinese government use of AI in Xinjiang, see the editorial column in the 28 December 2017 edition of the *Guardian*, 'The Guardian view on surveillance in China: Big Brother is watching'. For details about the use of facial-recognition glasses by Chinese police, see Kinling Lo's 7 February 2018 article in the *South China Morning Post* (International Edition), 'In China, these facial-recognition glasses are helping police to catch criminals'. Frank Tang's 2 June 2018 article in the same paper, 'China names 169 people banned from taking flights or trains under social credit system', contains details of how the social-credit system is being used to punish antisocial citizens. The quote from Nicholas Thompson and Ian Bremmer is from their 23 October 2018 *Wired* article, 'The AI Cold War that threatens us all'. The quote from an anonymous Chinese local appears in Tom Phillips's 18 January 2018 *Guardian* article, 'China testing facial-recognition surveillance system in Xinjiang'.

3.

For a good overview of the growth in AI research, and why this does not necessarily translate into the imminent arrival of an AGI, see Matt Turck's 18 April 2018 article on Hackermoon.com, 'Frontier AI: How far are we from artificial "general" intelligence, really?'

DeepMind's technical paper about using machine learning to play Atari games is online at arxiv.org/pdf/1312.5602v1.pdf. For OpenAI's account of training machines to play *DOTA* 2, see their blog at blog. openai.com/openai-five/, and for an account of the AI's defeat by pro players see James Vincent's 28 August 2018 story at *The Verge*, 'OpenAI's *DOTA 2* defeat is still a win for artificial intelligence'. For details about the difficulties DeepMind's AI had in playing *Montezuma's Revenge*, see 'Google's AI is now smart enough to play Atari like the pros' by Robert McMillan in the 25 February 2015 edition of *Wired*. For more on the Way of the Future, see Andrew J. Hawkins's 27 September 2017 article 'Anthony Levandowski, Uber's fired self-driving-car engineer, founded his own church', in *The Verge*.

4.

For a disturbing account of parasomnia, in which a sleeping person both drove a car and then killed, see the paper by Shreeya Popat and William Winslade in 24 April 2015 edition of *Neuroethics*, 'While you were sleepwalking: Science and neurobiology of sleep disorders & the enigma of legal responsibility of violence during parasomnia'. Douglas Coupland defines bambification on p. 48 of his 1991 novel *Generation X*. The quote from *Human Brain Function* (2013), by Richard Frackowiak et al., is from p. 269. I found the quote from *International Dictionary of Psychology* on p. 28 of the 23 June 2018 issue of *New Scientist*.

The 6 April 2017 edition of *New Scientist* has an account of how octopuses can edit their genes, 'Squid and octopus can edit and direct their own brain genes'. The quote from Peter Daws about Charles the octopus is taken from p. 54 of *Other Minds* by Peter Godfrey-Smith, and the quote from Stefan Linquist is from p. 56. The octopus fighting a shark appeared in episode 5 of *Blue Planet II*, 'Green Seas', which was broadcast on BBC1 on Sunday, 26 November 2017, and Craig Foster's quote is taken from the BBC website for the programme. The quote from Agustín Fuentes is taken from the very start of his book *The Creative Spark*, in the section 'Overture: Trumpeting creativity and a new synthesis'.

5.

The quotes from Peter Godfrey-Smith are from pp. 90–91 of his book *Other Minds*.

6.

Stuart Hameroff and Robert Penrose's paper regarding quantum vibrations inside microtubules, 'Consciousness in the universe: A review of the "Orch OR" theory', was published in *Physics of Life Reviews*, Volume 11, Issue 1, March 2014, pp. 39–78. Matthew Fisher's ideas about the nuclear spins of phosphorus atoms are discussed in his paper 'Quantum cognition: The possibility of processing with nuclear spins in the brain', published in the *Annals of Physics*, Volume 362, November 2015, pp. 593–602. Henry Markram's thoughts about seven-dimensional structures in the brain are discussed in Anil Ananthaswamy's 30 September 2017 *New Scientist* article, 'Throwing shapes', pp. 28–32.

7.

For an overview of the history of jetpacks and current developments in the field, see Dave Hall's 15 May 2018 *Guardian* article, 'Jetpacks: Why aren't we all flying to work?', which is also the source for the quotes from Isaac Asimov and Bill Suitor. The quote from Herbert Simon originally comes from his 1965 book *The Shape of Automation for Men and Management*, although I took it from p. 10 of *Surviving AI* by Calum Chace.

8.

For a report on the work by Michigan State University to train AI to recognise children from old photographs, see the 2 December 2017 story by Abigail Beall on p. 14 of the *New Scientist*, 'AI could match missing kids to old photos'.

3. PATTERNS

1.

For more details on Eric Meyer and his reaction to Facebook's 'Your Year in Review' feature, see his 29 December 2014 article on Slate.com, 'My year was tragic. Facebook ambushed me with a painful reminder'. Anna England Kerr wrote about her experience with Facebook in the 19 October 2018 *Huffington Post* article 'As I was grieving my stillborn daughter, Facebook was targeting me with baby adverts'.

2.

For more information on The Spirit of Gravity, see their website at spiritofgravity.com. For a discussion on how fewer songs are becoming hits, see Patrick Scott's 6 January 2017 *Daily Telegraph* article, 'With fewer number ones than ever before in 2016, has the success of streaming killed UK chart music?'

3.

The quotes from Yuval Noah Harari are from pp. 428 and 451 of his book *Homo Deus*. The statement by Susan Bidel that 50 per cent of data is assumed to be inaccurate comes from the 10 April 2018 BBC News report by Padraig Belton and Matthew Wall, 'More than 600 apps had access to my iPhone data'. For more information on how Google's picture-recognition AI classified black people as gorillas, see Loren Grush's 1 July 2015 article in *The Verge*, 'Google engineer apologizes after Photos app tags two black people as gorillas'. *The Verge* is also the source of a good – and delightfully titled – account of Microsoft's chatbot Tay, 'Twitter taught Microsoft's AI chatbot to be a racist asshole in less than a day', written by James Vincent on 24 March 2016.

4.

For more on the life of Timothy Leary and his metaphor of reality tunnels, see my 2006 book *I Have America Surrounded*. For more on the role of dopamine in confirmation bias, see 'Dopaminergic genes predict individual differences in susceptibility to confirmation bias' by Bradley B. Doll, Kent E. Hutchison and Michael J. Frank in the 20 April 2011 edition of the *Journal of Neuroscience*.

5.

For an account of the accuracy of facial-recognition systems, including their systematic biases, see 'Are face recognition systems accurate? Depends on your race', by Mike Orcutt, at technologyreview.com, dated 6 July 2016. For more about South Wales Police's use of the technology, see the 15 May 2018 BBC News article by Chris Foxx, 'Face recognition police tools "staggeringly inaccurate"', and for their use by the Met see the 15 May 2018 story 'Zero arrests, 2 correct matches, no criminals: London cops' facial recog tech slammed', by Rebecca Hill, in *The Register*. For an example of a mistake by Chinese facial recognition systems, see Melanie Ehrenkranz's 26 November 2018 *Gizmodo* story, 'Facial Recognition Flags Woman on Bus Ad for "Jaywalking" in China'.

For more on the death of Elaine Herzberg being down to a 'false positive', see 'Report: Software bug led to death in Uber's self-driving crash', written by Timothy B. Lee and posted on arstechnica.com on 7 May 2018. For details on how AI can be tricked into confusing a turtle with a gun, see James Vincent's 2 November 2017 report for *The Verge*, 'Google's AI thinks this turtle looks like a gun, which is a problem'. The 19 March 2018 report for *Motherboard* by Kaleigh Rogers, 'This hat can fool a face recognition system into thinking you're Moby', details how infrared light can be deployed to fool facial-recognition software. The 25 July 2017 *Guardian* article 'Google enters race for nuclear fusion technology', by Damian Carrington, discusses the use of AI as a tool.

Anyone interested in campaigning against giving AI responsibility for killing should visit www.stopkillerrobots.org.

6.

For more on the decline effect, see Jonah Lehrer's 13 December 2010 *New Yorker* article, 'The truth wears off: Is there something wrong with the scientific method?' The quote about Stanisław Lem by Simon Ings in the *New Scientist* was taken from his 19 November 2016 article, 'The man with the future inside him'. Lem's story about the machine that can make anything beginning with the letter 'n' is 'How the World was Saved', which appears in his 1965 story collection, *The Cyberiad*.

4. THE METAMODERN GENERATION

1.

Molly Ringwald's 6 April 2018 *New Yorker* article was 'What about "The Breakfast Club"? Revisiting the movies of my youth in the age of #MeToo'.

2.

St Paul's words come from 1 Corinthians 13:11. The statistic regarding the median age of first marriage for American women comes from p. 23 of Jean M. Twenge's *iGen*. Will Self's article 'The awful cult of the talentless hipster has taken over' appeared in the 15 September 2014 edition of the *New Statesman*, while Padraig Reidy's and Alex Rayner's *Guardian* article 'Is it OK to hate hipsters?' followed on 20 September 2014.

3.

Jean Twenge's words about the abrupt shift in teen behaviour are taken from p. 4 of her book *iGen*. That book is also the source for the statistic that 56 per cent more teens experienced a major depressive episode in 2015 compared to 2010 (p. 108) and that mobile-devices use is linked to unhappiness and mental-health issues if use exceeds more than two hours a day (p. 292). For details of the study that showed greater harm for girls than boys in excessive social media use, see the *Guardian* article 'Depression in girls linked to higher use of social media', written by Denis Campbell on 4 January 2019.

4.

The description of the data sets used by Jean Twenge is taken from page 9 of *iGen*. The statistic about British youth drinking less is reported by BBC News in the 10 October 2018 story 'Under-25s turning their backs on alcohol, study suggests'. That data from around the world matches the fall in sexual activity among young Americans is discussed in the December 2018 *Atlantic* article 'Why Are Young People Having So Little Sex?' by Kate Julian.

5.

For more information about the importance of play in human development, including details about the length of childhood in our australopithecine ancestors and Neanderthal cousins, see p. 190 of the *New Scientist* book *How to be Human*. Agustín Fuentes's words are taken from the 'Overture: Trumpeting creativity and a new synthesis' section of his book *The Creative Spark*. For more on the role of play in Finnish early education, see Patrick Butler's 20 September 2016 *Guardian* article, 'No grammar schools, lots of play: the secrets of Europe's top education system'.

For an overview of the Flynn effect, see Ronald Bailey's 1 June 2015 article on *Reason.com*, 'Average IQ scores have risen 30 points during the past 100 years'. Rory Smith's 14 June 2018 CNN report, 'IQ scores are falling and have been for decades, new study finds', covers evidence of the Flynn effect ending. The statistics regarding life expectancy are from a 17 December 2012 report by the Office of National Statistics, 'Mortality in England and Wales: Average life span, 2010'.

6.

Jean Twenge discusses whether safe spaces are fringe ideas on p. 155 of *iGen*, and how danger lies in speech on p. 156. The YouGov report detailing young people's unwillingness to talk to strangers is summarised by Matthew Smith in the 12 December 2017 article on yougov.co.uk, 'It's good to talk? Not if you're young or on public transport'. The origin of the *Beauty and the Beast* meme seems to be a 16 June 2017 tweet by @DanaSchwartzzz, while Janet Maslin's *New York Times* review of *Beauty and the Beast* was published on 13 November 1991. The 2015 study that examined IQ tests is discussed on p. 35 of the 21 July 2018 edition of *New Scientist*, in Linda Geddes's article 'The truth about intelligence'.

8.

The Chris Bateman quote comes from p. 133 of his 2018 book *The Virtuous Cyborg*. Eric Voegelin's description of the *metaxy* of Greek heroes is taken from p. 11 of *Metamodernism*, edited by Robin van den Akker, Alison Gibbons and Timotheus Vermeulen. Luke Turner's 2011 *Metamodernist Manifesto* is online at www.metamodernism.org, while his 12 January 2015 essay, 'Metamodernism: A brief introduction', can be found on www.metamodernism.com. The Stewart Lee quote is from his 2010 show *If You Prefer a Milder Comedian, Please Ask for One*. 'Will Gompertz reviews Childish Gambino's This is America video' can be found on the BBC News website, dated 12 May 2018. The quote from Robin van den Akker and Timotheus Vermeulen is on p. 5 of *Metamodernism*. The Rana Dasgupta quote is taken from his 5 April 2018 *Guardian* article, 'The demise of the nation state'.

10.

Jay Owens's essay 'The age of post-authenticity and the ironic truths of meme culture' was posted on Medium.com on 11 April 2018. Samantha Fuentes's opinion of President Trump is quoted in Julie Hirschfeld Davis's 22 February 2018 *New York Times* article, 'What do jotted talking points say about Trump's empathy?' Footage of speech at the March for Our Lives event in Washington is available on the CNN website, dated 24 March 2018 and entitled 'Parkland survivor vomits during gun speech'.

5. THE DREAM OF SPACE

1.

Video footage of the 6 February 2018 press conference in which Elon Musk tells reporters that he is 'tripping balls' is on the CBS News You-Tube channel under the title 'Elon Musk celebrates successful Falcon Heavy rocket launch'. Naomi Klein's tweet can be found at twitter.com/ NaomiAKlein/status/961000123609296902 (Warren Ellis's tweet has since been deleted, along with much of his Twitter history).

2.

Gene Roddenberry's response of 'Fuck off, then' is recounted by his son, Rod Roddenberry, on p. 36 of *The Fifty-Year Mission: The First 25 Years*, by Edward Gross and Mark A. Altman, and his comments about avoiding censorship appear on p. 66. The same book is also the source of quotes from Fred Freiberger (p. 201), Ande Richardson (p. 133) and Ed Naha (p. 37). The episode in which Captain Kirk is trapped in the body of a woman is 'Turnabout Intruder', series 3, episode 24, which originally aired on 3 June 1969.

3.

The quotes from Nicholas Meyer are taken from p. 409 of *The Fifty-Year Mission: The First 25 Years* by Edward Gross and Mark A. Altman, and the quote from Ralph Winter is from p. 435. The statistics about smoking levels in the United States are from the tobacco fact sheets section of the Center for Disease Control and Prevention website.

4.

The quote from Hans Beimler is taken from *The Fifty-Year Mission: The Next 25 Years* by Edward Gross and Mark A. Altman, ebook location 2557. David Gerrold's quote is from location 1769 and Herb Wright's quote is from location 1795 in the same book. The *Star Trek: The Next Generation* episode in which Q flings the starship *Enterprise* far across the galaxy is 'Q Who', season 2, episode 16, first broadcast on 8 May 1989. The season

4, episode 2, story 'Family', which was broadcast on 1 October 1990, shows Picard fighting his brother.

5.

The quote from Simon Pegg is from 'Simon Pegg: Boldly going nerd', by Erik Adams, a 9 June 2011 article on avclub.com.

6.

The Douglas Adams quote is from his 1979 novel *The Hitchhiker's Guide to the Galaxy*. The quotes from Kim Stanley Robinson's *Aurora* are taken from pp. 208–9, 178 and 428. Kameron Hurley describes her novel *The Stars Are Legion* as a 'lesbians-in-space organic worldship space opera' on her Patreon page, at www.patreon.com/kameronhurley.

7.

Captain Lorca namechecks Elon Musk in *Star Trek: Discovery* season 1, episode 4, 'The Butcher's Knife Cares Not for the Lamb's Cry', broadcast on 8 October 2017. For details of the UK government's plans to switch to electric vehicles, see the 26 July 2017 story 'New diesel and petrol vehicles to be banned from 2040 in UK' on the BBC News website. Jaguar Land Rover's and Volvo's plans to stop producing non-electric vehicles are discussed in Adam Vaughan's 7 September 2017 *Guardian* article, 'Jaguar Land Rover to make only electric or hybrid cars from 2020'. The European Union's attitudes to electric vehicles are discussed in the 8 November 2017 Bloomberg article 'It's not Tesla that's pushing Europe in electric-car race', by Ewa Krukowska, Jonathan Stearns and Nikos Chrysoloras. The *Wall Street Journal* review of the Tesla 3, 'First test drive of the Tesla Model 3 performance: A thrilling, modern marvel', by Dan Neil, was published on 19 July 2018.

For more information on the cost of launching rockets, see qz.com's 4 December 2013 article 'SpaceX just made rocket launches affordable. Here's how it could make them downright cheap', by Tim Fernholz, and spacenews.com's 25 April 2016 story 'SpaceX's reusable Falcon 9: What are the real cost savings for customers?', by Peter B. de Selding.

8.

For more on the impact that Mars's gravity would have on us biologically, see Kevin Fong's 11 February 2014 *Wired* article, 'The strange, deadly effects Mars would have on your body'. The problems of radiation are discussed in Sarah Frazier's 30 September 2015 NASA.gov article, 'Real Martians: How to protect astronauts from space radiation on Mars'. Elon Musk's comments about building an underground Martian city are taken from Mark Kaufman's 20 July 2017 article for Inverse.com, 'Elon Musk's L.A. tunnel boring project is just practice for Mars'. The lack of sufficient carbon dioxide on Mars in order to terraform the planet is discussed in Leah Crane's 4 August 2018 *New Scientist* story, 'Plan to terraform Mars is out of gas'.

6. BETWEEN THE REAL AND THE VIRTUAL

1.

For more on the idea that VR is an empathy machine, see Techcrunch's 2014 article 'Virtual reality, the empathy machine', by Josh Constine, or Chris Milk's 2015 TED talk, 'How virtual reality can create the ultimate empathy machine'. The quote from Jaron Lanier comes from p. 60 of his 2017 book, *Dawn of the New Everything*.

2.

For a good description on how 'redpilling' works, see Robert Evans's 11 October 2018 essay 'From memes to infowars: How 75 fascist activists were "red-pilled"', on www.bellingcat.com. For more on the political side of the *Star Wars: The Last Jedi* backlash, see the 2 October 2018 *Guardian* article '*Star Wars: The Last Jedi* abuse blamed on Russian trolls and "political agendas"', by Andrew Pulver. The backlash against Oculus is detailed in the 26 March 2014 *Time* magazine article 'When crowdfunding goes corporate: Kickstarter backers vent over Facebook's Oculus buy', by Victor Luckerson. Palmer Luckey's political donations are discussed in the 23 September 2016 *Guardian* article 'Oculus Rift founder Palmer Luckey spends fortune backing pro-Trump "shitposts"', by Alex Hern. The subject of Oculus supplying user data to Facebook is the focus of Adi

Robertson's 9 April 2018 report for *The Verge*, 'How much VR user data is Oculus giving to Facebook?' Oculus's privacy policy can be viewed online at www.oculus.com/legal/privacy-policy. Mark Zuckerberg is quoted as saying that Facebook having an influence on the 2016 American election was a 'pretty crazy idea' in Aarti Shahani's 11 November 2016 report for npr.org, 'Zuckerberg denies fake news on Facebook had impact on the election'. The quote from Jaron Lanier is taken from his 2018 nymag.com interview with Noah Kulwin, 'One has this feeling of having contributed to something that's gone very wrong'.

3.

The Fukushima VR video is dated 12 March 2018 and can be found on the *Guardian* under the heading 'Fukushima 360: walk through a ghost town in the nuclear disaster zone – video'. Steven Pinker's statistics about nuclear safety appear on p. 147 of his book *Enlightenment Now*. The George Monbiot pro-nuclear quote appears in his 21 March 2011 *Guardian* column, 'Why Fukushima made me stop worrying and love nuclear power'. For more on the embrace of nuclear by the Union of Concerned Scientists, see the 24 November 2018 *New Scientist* article 'Nuclear power play' by Mark Lynas. We know of Socrates' suspicion of the written word thanks to Plato's dialogue *The Phaedrus*.

4.

For more about the psychological impact of being Einstein in VR, see *Science Daily's* 9 July 2018 story 'Seeing yourself as Einstein may change the way you think'. 'VR can ease fear of public speaking', a *New Scientist* story from 13 October 2018, discusses research into easing phobias. The quote from Castronova was taken from p. 237 of Chris Bateman's *Imaginary Games*.

7. PSYCHIC POLLUTION

1.

The 1 November 2017 American Psychological Association survey was called 'APA stress in America survey: US at "lowest point we can remember"; future of nation most commonly reported source of stress'. The 3 May 2012 Fairleigh Dickinson University report, 'What you know depends on what you watch: Current events knowledge across popular news sources', can be found online at publicmind.fdu.edu/2012/ confirmed. For more on the impact negative news can have on personal worries, see 'The psychological impact of negative TV news bulletins: The catastrophizing of personal worries', by Wendy M. Johnston and Graham C. L. Davey, in the 13 April 2011 issue of the *British Journal of Psychology*.

2.

For an overview of the evolutionary advantages of social interaction, see Pascal Vrticka's 16 November 2013 article on Huffingtonpost.com, 'Evolution of the "Social Brain" in Humans: What Are the Benefits and Costs of Belonging to a Social Species?' The Steven Pinker quote is taken from pp. 40–41 of his book *Enlightenment Now*. The website of Dopamine Labs is usedopamine.com, and more details about the company and their founder, Ramsay Brown, can be found in the 2017 Techcrunch article 'Meet the tech company that wants to make you even more addicted to your phone', by Jonathan Shieber. Ramsay Brown's *60 Minutes* interview was uploaded to the Boundless Minds YouTube channel on 13 April 2017 under the title *Dopamine Labs: 60 Minutes*. The Ginsberg-paraphrasing quote from Jeff Hammerbacher appears on p. 279 of Kevin Kelly's book *The Inevitable*.

3.

For more about Stanford's Persuasive Tech Lab, see its website, http://captology.stanford.edu. The quotes from Sean Parker are taken from Alex Hern's 23 January 2018 *Guardian* article, ' "Never get high on your own supply" – why social media bosses don't use social media'. Jaron Lanier's Vox podcast can be found on vox.com under the heading 'You will love this conversation with Jaron Lanier, but I can't describe it', dated 16 January 2018.

The quotes from Chamath Palihapitiya are from James Vincent's 11 December 2017 report in *The Verge*, 'Former Facebook exec says social media is ripping apart society'. As an example of Roger McNamee speaking out about Facebook, see his 13 January 2018 *Guardian* article, 'I was

Mark Zuckerberg's mentor. Today I would tell him: your users are in peril'. For more on the Center for Humane Technology, see its website, humanetech.com. Tristan Harris's comments about Snapchat streaks are taken from 'Ex-tech workers plead with Facebook: Consider the harm you're doing to kids', an 8 February 2018 *Guardian* article by David Smith.

Steve Jobs is quoted about not letting his kids use an iPad in Nick Bilton's 10 September 2014 *New York Times* story, 'Steve Jobs was a low-tech parent', while the quote from Tim Cook is taken from the *Guardian* story 'Apple's Tim Cook: "I don't want my nephew on a social network"', by Samuel Gibbs on 19 January 2018. The quote from an R. J. Reynolds executive appears at www.ashscotland.org.uk/about-us/job-done. For details about the upset caused by McDonald's advertising, see the 17 May 2017 *Daily Telegraph* report 'McDonald's pulls "exploitative" grieving child advert from screens'.

4.

For a discussion on the link between screen time and happiness, see Jean Twenge's 26 January 2018 article on qz.com, 'Most unhappy people are unhappy for the exact same reason'. The Justin Rosenstein quote is taken from Douglas Heaven's article 'Techlash', on p. 31 of the 10 February 2018 edition of *New Scientist*. For more on the role of D1 and D2 dopamine receptors in addiction, see the 3 February 2018 *New Scientist* story 'A helping hand', by Alice Klein, on p. 39. For more on the negative impact of oxytocin, see 'Oxytocin ain't behavin' how scientists thought it would', by Christopher Bergland in the 23 September 2017 *Psychology Today*. Research that links glutamate and the neurotransmitter GABA to smartphone addiction is discussed on p. 16 of the 9 December 2017 issue of *New Scientist*, in the article 'Tech addiction's chemical cocktail'.

5.

Mark Zuckerberg's use of the phrase 'dumb fucks' to describe people who trusted him was reported by *Business Insider* in a 13 May 2010 story by Nicholas Carlson, 'Well, these new Zuckerberg IMs won't help Facebook's privacy problems'. The Cornell University report on emotional contagion is discussed in Greg Taylor's short 2018 ebook *Weaponizing Facebook* at location 24, and the Australian memo is discussed at location 67 in the same ebook. The Buzzfeed report on how fake news was more successful on Facebook than real news was 'This analysis shows how viral fake election news stories outperformed real news on Facebook', by Craig Silverman on 16 November 2016.

The Channel 4 investigation into Cambridge Analytica, including the quotes from Mark Turnbull, are from their 20 March 2018 story 'Exposed: Undercover secrets of Trump's data firm'. George Soros's remarks are taken from a 26 January 2018 *Guardian* story by Olivia Solon, 'George Soros: Facebook and Google a menace to society'. Solon also wrote the 1 June 2018 *Guardian* story 'Teens are abandoning Facebook in dramatic numbers, study finds'. For details of Facebook's share price collapse in July 2018, see marketewatch.com's 26 July 2018 story 'Facebook stock drops roughly 20 per cent, loses $120 billion in value after warning that revenue growth will take a hit', by Max A. Cherney.

6.

For an update on the state of the ozone layer, see Sarah Knapton's 5 January 2018 *Daily Telegraph* story, 'Hole in ozone layer has shrunk thanks to worldwide ban of CFCs, Nasa confirms'.

8. FIXING THINGS

1.

The Andy Coghlan quote about wildfires is taken from his 21 April 2018 report for the *New Scientist*, 'Wildfires created vast thunderstorm', on p. 14. The VR film of the fires was *This is Climate Change: Fire*, by WITHIN.

2.

The quotes from *Uncivilisation* by Paul Kingsnorth and Dougald Hine are taken from pp. 7 and 5–6. The Dark Mountain website is at https://dark-mountain.net. The *New Scientist* editorial quoted was on p. 3 of the 28 July 2018 issue, and called 'The state of life on Earth'. The quotes from Paul Kingsnorth's wider writings are all collected in his book *Confessions of a Recovering Environmentalist* and can be found on pp. 24 and 94–5. The claim that the amount of the Earth's biomass has fallen by half since the beginning of human civilisation is found in Michael Le Page's 26 May 2018 story at p. 8 of the *New Scientist*, 'Half of Earth's life has vanished'.

3.

'Nature's collapse threatens society', a report on the Intergovernmental Platform on Biodiversity and Ecosystem services, can be found on p. 6 of the 31 March 2018 *New Scientist*. The *Proceedings of the National Academy of Sciences* report on biodiversity is discussed in 'Earth's sixth mass extinction event under way, scientists warn', a 10 July 2017 *Guardian* article by Damian Carrington. The quote from Michael McCarthy is taken from his 26 March 2018 *Guardian* article, 'Britain has lost half its wildlife. Now's the time to shout about it'. For more on predictions of soil loss, see Chris Arsenault's 5 December 2014 *Scientific American* article, 'Only 60 years of farming left if soil degradation continues'. For more on UN population predictions, see the 31 July 2015 *National Geographic* story 'World population expected to reach 9.7 billion by 2050', by Rachel Becker.

4.

The quotes from E. O. Wilson are taken from pp. 2 and 3 of his book *Half-Earth*. The quotes from James Watson and Jonathan Baillie are taken from Michael Le Page's 22 September 2018 *New Scientist* article, 'Half the planet should be set aside for wildlife – to save ourselves'. This article also discusses the Convention on Biological Diversity targets. The Rewilding Europe report is discussed on pp. 154–5 of Isabella Tree's book *Wilding*.

For more on the Columbian rainforest, see the 2 July 2018 WWF story 'Colombia's Serranía de Chiribiquete is now the world's largest tropical rainforest national park'. For details of the Seychelles' marine protected areas, see Elaina Zachos's 21 February 2018 *National Geographic* article, 'Ocean refuge the size of Great Britain announced'. For more on the increase in trees in the UK, see 'England's woodlands growing to 1,000-year record total', a 22 November 2001 *Guardian* article by Paul Brown. Fred Pearce's 19 May 2018 *New Scientist* article, 'The hidden international trade in deforestation', discussed the reduction in tree numbers in poorer countries. E. O. Wilson's description of mankind is taken from p. 1 of his book *Half-Earth*.

5.

The quote from Isabella Tree is taken from p. 308 of her book *Wilding*. For more on controversies surrounding a Dutch rewilding scheme, see Patrick Barkham's 27 April 2018 *Guardian* article, 'Dutch rewilding experiment sparks backlash as thousands of animals starve'. For more on the monitoring of Borneo rainforest by AI, see Richard Kemeny's p. 10 story in the 6 October 2018 *New Scientist*, 'AI eavesdrops on Borneo's rainforests'. The quotes from E. O. Wilson and Alexander von Humboldt are both from p. 78 of Wilson's book *Half-Earth*.

6.

The statistics about the amount of stuff used per person in the UK are taken from p. 136 of Steven Pinker's book *Enlightenment Now*.

7.

Nick Boles MP's comments are discussed in Heather Stewart's 28 December 2017 *Guardian* article, 'Tory MP condemns universal basic income "on moral grounds"'. The Buckminster Fuller quote is from the 30 March 1970 edition of *New York* magazine. The quotes from Viktor E. Frankl are taken from pp. 82, 83 and 85 of his book *Man's Search for Meaning*, while the quote from Nietzsche originated in his 1889 book *Twilight of the Idols*.

The YouGov survey mentioned is discussed on the yougov.co.uk website in Will Dahlgreen's 12 August 2015 article, '37% of British workers think their jobs are meaningless'. The 2013 Harvard Business Review survey is discussed on the World Economic Forum website, in a 12 April 2017 article by Rutger Bregman called 'A growing number of people think their job is useless. Time to rethink the meaning of work'. David Graeber discusses 'bullshit jobs' in his 2018 book of the same name. Details of the University of Manchester study are on their website under the title 'Having a bad job can be worse for your health than being unemployed', dated 11 August 2017. William Blake's notes on Sir Joshua Reynolds's opinions are noted on p. 61 of Peter Ackroyd's biography *Blake*. Mark Carney's words are from his keynote talk at the Public Policy Forum in April 2018, which can be seen on YouTube (36 minutes in).

8.

For a report on the use of the legal system to fight climate change, including the case of the 'climate kids', see the 18 August 2018 *New Scientist* article by Fred Pearce, 'Polluter pays?'. Greta Thunberg's December 2018 address to the U.N. plenary session in Poland can be viewed on the *Democracy Now!* YouTube channel, entitled 'You Are Stealing Our Future: Greta Thunberg, 15, Condemns the World's Inaction on Climate Change'. For further details about the International Congress of Youth Voices, see their website at www.internationalcongressofyouthvoices.com. The survey funded by the Bill & Melinda Gates Foundation is discussed in the 24 September 2018 *Guardian* story '"Our time is now": world youth poll reveals unexpected optimism', by Rebecca Ratcliffe and Carmen Aguilar. The account of Curt Richter's 1957 experiment is taken from a 13 October 2018 *New Scientist* interview with the Argentinian artist Tomás Saraceno, 'Why Saraceno's balloon art is more than hot air', written by Simon Ings.

9. MORE THAN INDIVIDUAL

1.

For more on the UAE's City of Wisdom, see Philip Perry's 14 March 2017 story on bigthink.com, 'Forget colonization. The UAE plans to build a city on Mars by 2117'. *The Habitat* podcast is produced by Gimlet Media.

2.

John Doran's 24 November 2017 article 'Welcome to the naked, paint-dripping world of New Weird Britain' appeared in *VICE* magazine and *Noisey's 2017 Music Issue*. The Hakim Bey quotes are taken from pp. 7, 10 and 44 of his book *Immediatism*.

3.

The quotes from Henry Timms and Jeremy Heimans are from pp. 2 and 1 of their book *New Power*. Chris Bateman discusses the scale of our networks on p. 77 of *The Virtuous Cyborg*. The quote from Basar, Coupland

and Obrist's book *The Age of Earthquakes* is taken from pp. 155–7. The quote from St Paul is at Colossians 3:2.

4.

The quote from Rachel Botsman is taken from the introduction to her 2017 book *Who Can You Trust?: How Technology Brought Us Together – and Why It Could Drive Us Apart*. The Carl Jung quote is from his autobiographical 1961 book *Memories, Dreams, Reflections*. The quote from Rebecca Adams is taken from Alex Williams's *New York Times* article 'Friends of a certain age', from 13 July 2012. For more on the *hikikomori*, see Michael Hoffman's 9 October 2011 *Japan Times* report, 'Nonprofits in Japan help "shut-ins" get out into the open'.

5.

For a full report on the city hipster index, see Frederick O'Brien's 19 April 2018 report on movehub.com, 'The Hipster Index: Brighton pips Portland to global top spot'. For details of the survey declaring Brighton to be the happiest place in the UK, see the 30 May 2008 Sky News report 'Brighton: The happiest place in the UK'. I am indebted to Mr Adam King of the YouTubes for details of the world's first adult movie. For more on Brighton's entrepreneurial spirit, see the 25 November 2018 story by Peter Lindsey in the *Brighton Argus*, 'Brighton is officially the best city to start a small business in UK'. The quote from Jeremy Lent is taken from p. 15 of his 2017 book *The Patterning Instinct*. For more on Amazon's plans to adapt the work of Iain M. Banks, see Andrew Liptak's 21 February 2018 report for *The Verge*, 'Amazon is developing a series based on Iain M. Banks' sci-fi novel *Consider Phlebas*'.

BIBLIOGRAPHY

Banks, Iain M., *The Player of Games* (Macmillan, 1988)

——, *Use of Weapons* (Orbit, 1990)

Bartlett, Jamie, *The Dark Net* (Windmill Books, 2015)

Basar, Shumon, Douglas Coupland and Hans Ulrich Obrist, *The Age of Earthquakes: A Guide to the Extreme Present* (Penguin Books, 2015)

Bateman, Chris, *Chaos Ethics* (Zero Books, 2014)

——, *Imaginary Games* (Zero Books, 2011)

——, *The Virtuous Cyborg* (Eyewear Publishing, 2018)

Bey, Hakim, *Immediatism* (AK Press, 1994)

Booker, Christopher, *The Seven Basic Plots: Why We Tell Stories* (Continuum, 2004)

Chace, Calum, *Surviving AI: The Promise and Peril of Artificial Intelligence* (Three Cs Publishing, 2015)

Cline, Ernest, *Ready Player One* (Penguin Books, 2011)

Coupland, Douglas, *Generation X: Tales from an Accelerated Culture* (Abacus, 1992)

Ellis, Warren, *Normal* (Farrar, Straus and Giroux, 2016)

BIBLIOGRAPHY

Evans, Alex, *The Myth Gap: What Happens When Evidence and Arguments aren't Enough* (Transworld Digital, 2017)

Foer, Franklin, *World Without Mind: The Existential Threat of Big Tech* (Jonathan Cape, 2017)

Frankl, Viktor E., *Man's Search for Meaning* (Rider, 1959)

Fuentes, Agustín, *The Creative Spark: How Imagination Made Humans Exceptional* (Dutton, 2017)

Godfrey-Smith, Peter, *Other Minds: The Octopus and the Evolution of Intelligent Life* (William Collins, 2017)

Greenfield, Adam, *Radical Technologies: The Design of Everyday Life* (Verso, 2017)

Gross, Edward and Mark A. Altman, *The Fifty-Year Mission: The Complete, Uncensored, Unauthorized Oral History of* Star Trek: *The First 25 Years* (Thomas Dunne Books, 2016)

—— and ——, *The Fifty-Year Mission: The Next 25 Years: From The Next Generation to J. J. Abrams: The Complete, Uncensored, and Unauthorized Oral History of* Star Trek (Thomas Dunne Books, 2016)

Harari, Yuval Noah, *Homo Deus: A Brief History of Tomorrow* (Vintage, 2017)

——, *Sapiens: A Brief History of Humankind* (Vintage, 2011)

Higgs, John, *I Have America Surrounded: The Life of Timothy Leary* (Friday Project, 2006)

——, *Stranger Than We Can Imagine: Making Sense of the Twentieth Century* (Weidenfeld & Nicolson, 2015)

Hurley, Kameron, *The Stars Are Legion* (Angry Robot, 2017)

Huxley, Aldous, *Brave New World* (Chatto & Windus, 1932)

Kelly, Kevin, *The Inevitable: Understanding the 12 Technological Forces That Will Shape Our Future* (Penguin Books, 2016)

Kingsnorth, Paul, *Confessions of a Recovering Environmentalist* (Faber & Faber, 2017)

——, *The Wake* (Unbound, 2014)

——, and Dougald Hine, *Uncivilisation: The Dark Mountain Manifesto* (The Dark Mountain Project, 2009)

Krauss, Lawrence M., *The Physics of Star Trek* (Flamingo, 1997)

Lanier, Jaron, *The Dawn of the New Everything: A Journey through Virtual Reality* (The Bodley Head, 2017)

Lawton, Graham, and Jeremy Webb (*New Scientist*), *How to be Human* (John Murray, 2017)

Leary, Timothy, *Chaos and Cyber Culture* (Ronin Publishing, 1994)

Lem, Stanisław, *The Cyberiad* (Seabury Press, 1974)

——, *Solaris* (Faber & Faber, 1961)

Michell, John, *The John Michell Reader: Writings and Rants of a Radical Traditionalist* (Inner Traditions, 2015)

Parfrey, Adam (ed.), *Apocalypse Culture* (Feral House, 1990)

Parkin, Lance, *The Impossible Has Happened: The Life and Work of Gene Roddenberry, Creator of* Star Trek (Aurum Press, 2016)

Pinker, Steven, *Enlightenment Now: The Case for Reason, Science, Humanism and Progress* (Allen Lane, 2018)

Robinson, Kim Stanley, *Aurora* (Orbit, 2015)

——, *Red Mars* (HarperVoyager, 1992)

Rushkoff, Douglas, *Throwing Rocks at the Google Bus: How Growth Became the Enemy of Prosperity* (Penguin Books, 2016)

Segal, Lynne, *Radical Happiness: Moments of Collective Joy* (Verso, 2017)

Shriver, Lionel, *The Mandibles: A Family, 2029–2047* (The Borough Press, 2016)

Strang, Em, Nick Hunt and Cate Chapman (eds.), *Dark Mountain Issue 10: Uncivilised Poetics* (The Dark Mountain Project, 2016)

Taylor, Greg, *Weaponizing Facebook: How Companies Like Cambridge Analytica are Manipulating Your Mind, and What You Can Do about It* (Daily Grail Publishing, 2018)

Timms, Henry and Jeremy Heimans, *New Power: How It's*

Changing the 21st Century – and Why You Need to Know
(Macmillan, 2018)

Tree, Isabella, *Wilding: The Return of Nature to a British Farm*
(Picador, 2018)

Twenge, Jean M., *Generation Me: Why Today's Young Americans are
More Confident, Assertive, Entitled – and More Miserable Than
Ever Before* (Atria Paperback, 2006)

——, *iGen: Why Today's Super-Connected Kids are Growing Up
Less Rebellious, More Tolerant, Less Happy – and Completely
Unprepared for Adulthood* (Atria Books, 2017)

Vance, Ashlee, *Elon Musk: How the Billionaire CEO of SpaceX and
Tesla is Shaping Our Future* (Virgin Books, 2016)

Van den Akker, Robin, Alison Gibbons and Timotheus
Vermeulen (eds.), *Metamodernism: Historicity, Affect and Depth
after Postmodernism* (Rowman and Littlefield International,
2017)

Warwick, Kevin, *Artificial Intelligence: The Basics* (Routledge, 2012)

Wilson, Edward O., *Half-Earth: Our Planet's Fight for Life*
(Liveright, 2016)

ACKNOWLEDGEMENTS

Huge thanks, big love and a bright future to all who have helped with this book. In data-friendly alphabetical order: Jason Arnopp, Sarah Ballard, David Bramwell, Joe Brown, Daisy Campbell, Eric Drass (as if that's a real name), Alistair Fruish, Melinda Gebbie, Holly Harley, Eli Keren, Joanne Mallon, Scott McPherson, Paul Murphy, Michelle Olley, Andrew O'Neill, Matt Pearson and Virginia Woolstencroft.

INDEX